Fast Quantitative Magnetic Resonance Imaging

Synthesis Lectures on Biomedical Engineering

Editor

John D. Enderle, *University of Connecticut*

Lectures in Biomedical Engineering will be comprised of 75- to 150-page publications on advanced and state-of-the-art topics that span the field of biomedical engineering, from the atom and molecule to large diagnostic equipment. Each lecture covers, for that topic, the fundamental principles in a unified manner, develops underlying concepts needed for sequential material, and progresses to more advanced topics. Computer software and multimedia, when appropriate and available, are included for simulation, computation, visualization and design. The authors selected to write the lectures are leading experts on the subject who have extensive background in theory, application and design.

The series is designed to meet the demands of the 21st century technology and the rapid advancements in the all-encompassing field of biomedical engineering that includes biochemical processes, biomaterials, biomechanics, bioinstrumentation, physiological modeling, biosignal processing, bioinformatics, biocomplexity, medical and molecular imaging, rehabilitation engineering, biomimetic nano-electrokinetics, biosensors, biotechnology, clinical engineering, biomedical devices, drug discovery and delivery systems, tissue engineering, proteomics, functional genomics, and molecular and cellular engineering.

Fast Quantitative Magnetic Resonance Imaging
Guido Buonincontri, Joshua Kaggie, and Martin Graves
December 2019

3D Electro Rotation of Single Cells
Liang Huang and Wenhui Wang
November 2019

Spatiotemporal Modeling of Influenza: Partial Differential Equation Analysis in R
William E. Schiesser
May 2019

PDE Models for Atherosclerosis Computer Implementation in R
William E. Schiesser
November 2018

Sensory Organ Replacement and Repair
Gerald E. Miller
2006

Artificial Organs
Gerald E. Miller
2006

Signal Processing of Random Physiological Signals
Charles S. Lessard
2006

Image and Signal Processing for Networked E-Health Applications
Ilias G. Maglogiannis, Kostas Karpouzis, and Manolis Wallace
2006

Fast Quantitative Magnetic Resonance Imaging
Guido Buonincontri, Joshua Kaggie, and Martin Graves

ISBN: 978-3-031-00539-8 paperback
ISBN: 978-3-031-01667-7 eBook
ISBN: 978-3-031-00046-1 hardcover

DOI 10.1007/978-3-031-01667-7

A Publication in the Springer Nature series
SYNTHESIS LECTURES ON ADVANCES IN AUTOMOTIVE TECHNOLOGY

Lecture #59
Series Editor: John D. Enderle, University of Connecticut

Series ISSN 1930-0328 Print 1930-0336 Electronic

Fast Quantitative Magnetic Resonance Imaging

Guido Buonincontri
Imago 7 Foundation

Joshua Kaggie and Martin Graves
University of Cambridge

SYNTHESIS LECTURES ON BIOMEDICAL ENGINEERING #59

ABSTRACT

Among medical imaging modalities, magnetic resonance imaging (MRI) stands out for its excellent soft-tissue contrast, anatomical detail, and high sensitivity for disease detection. However, as proven by the continuous and vast effort to develop new MRI techniques, limitations and open challenges remain. The primary source of contrast in MRI images are the various relaxation parameters associated with the nuclear magnetic resonance (NMR) phenomena upon which MRI is based. Although it is possible to quantify these relaxation parameters (qMRI) they are rarely used in the clinic, and radiological interpretation of images is primarily based upon images that are relaxation time weighted. The clinical adoption of qMRI is mainly limited by the long acquisition times required to quantify each relaxation parameter as well as questions around their accuracy and reliability. More specifically, the main limitations of qMRI methods have been the difficulty in dealing with the high inter-parameter correlations and a high sensitivity to MRI system imperfections.

Recently, new methods for rapid qMRI have been proposed. The multi-parametric models at the heart of these techniques have the main advantage of accounting for the correlations between the parameters of interest as well as system imperfections. This holistic view on the MR signal makes it possible to regress many individual parameters at once, potentially with a higher accuracy. Novel, accurate techniques promise a fast estimation of relevant MRI quantities, including but not limited to longitudinal (T_1) and transverse (T_2) relaxation times. Among these emerging methods, MR Fingerprinting (MRF), synthetic MR (syMRI or MAGIC), and T_1–T_2 Shuffling are making their way into the clinical world at a very fast pace. However, the main underlying assumptions and algorithms used are sometimes different from those found in the conventional MRI literature, and can be elusive at times. In this book, we take the opportunity to study and describe the main assumptions, theoretical background, and methods that are the basis of these emerging techniques.

Quantitative transient state imaging provides an incredible, transformative opportunity for MRI. There is huge potential to further extend the physics, in conjunction with the underlying physiology, toward a better theoretical description of the underlying models, their application, and evaluation to improve the assessment of disease and treatment efficacy.

KEYWORDS

spatial encoding, contrast encoding, spatial decoding, contrast decoding

Contents

CHAPTER 1

Introduction

In this chapter we introduce the basic principles of nuclear magnetic resonance (MR), including nuclear spin and polarization in a static magnetic field. The Bloch equation is then introduced to describe the response of the macroscopic magnetization to a radiofrequency (RF) pulse and its subsequent relaxation. We then provide a brief overview of the hardware associated with a magnetic resonance imaging system (MRI).

1.1 NUCLEAR MAGNETIC RESONANCE

The phenomenon of nuclear magnetic resonance (NMR) was first demonstrated independently and virtually simultaneously by Bloch [1-1] working at Stanford University and Purcell, Torrey, and Pound [1-2] working at Harvard University. The impact of their work was immediate, and the applications of NMR have steadily widened from physics and chemistry to biology and medicine. Bloch and Purcell were subsequently jointly awarded the 1952 Nobel Prize for physics *"for their development of new methods for nuclear magnetic precision measurements and discoveries in connection therewith."*

1.2 BASIC CONCEPTS

It has been known since the 1920s that certain atomic nuclei, which are made up of protons and neutrons (nucleons), possess an intrinsic angular momentum (J) or spin—like a gyroscope spinning on its axis. In most atomic nuclei, nucleons of the same type rotate in opposite directions and are paired together so that their total angular momentum or net spin is zero. The magnitude and direction of this angular momentum is characterized by a nuclear spin quantum number, I, to zero, half-integer and integer values. Those nuclei for which $I = 0$ do not possess spin angular momentum and therefore do not exhibit magnetic resonance phenomena. The nuclei of ^{12}C and ^{16}O fall into this category. Nuclei for which $I = \frac{1}{2}$ include ^{1}H, ^{19}F, ^{13}C, ^{31}P, and ^{15}N, while ^{2}H and ^{14}N have $I = 1$. The angular momentum is quantized as follows:

$$J = \hbar \left[I(I+1) \right]^{\frac{1}{2}}, \qquad (1.1)$$

where $\hbar$ is Planck's constant divided by 2π. The component of J in the z direction is given by $J_z = \hbar m_I$ where m_I is the magnetic quantum number and has values of $-I \leq m_I \leq I$, for a total of $2I + 1$ values. Therefore, for $I = \frac{1}{2}$ there are two possible orientations with $m_I = \frac{1}{2}$ or $-\frac{1}{2}$ (Figure 1.1).

Since the nucleus is electrically charged, one result of this spin is that the charge rotates as well and an electric current flows about the axis of rotation. This current generates a magnetic field and each nucleus of this type has a magnetic moment or dipole (μ) associated with it and can be considered to act as a tiny bar magnet. Fortunately, the single proton nucleus of hydrogen, the most abundant element in the body, possesses this property and is consequently the major nucleus of interest in MRI. Other nuclei in the body also exhibit this property but their lower abundance and sensitivity to the experiment make their investigation more difficult. The magnetic moment is given by

$$\boldsymbol{\mu} = \gamma \boldsymbol{J}, \tag{1.2}$$

where γ is the gyromagnetic ratio, a fundamental constant of the nucleus.

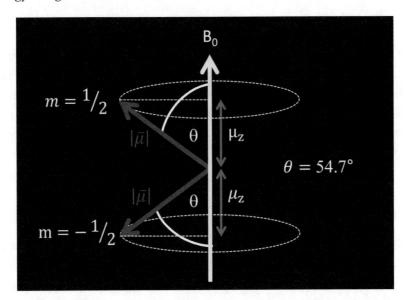

Figure 1.1: The two orientations of the nuclear spin angular momentum $\boldsymbol{J}$ for a nucleus with $I = \frac{1}{2}$.

Therefore, in the presence of a static magnetic field, the magnetic moments precess with two possible orientations either parallel or anti-parallel to the static magnetic field. The energy difference between the two states in the presence of a static magnetic field $\boldsymbol{B}$ is

$$\epsilon = \boldsymbol{\mu} \cdot \boldsymbol{B} = \gamma \hbar m_I B. \tag{1.3}$$

Selection rules only allow transitions between $m_I = -I, -I + 1, \ldots, +I$ so for protons

$$\Delta \epsilon = (\frac{1}{2} - -\frac{1}{2}) \gamma \hbar B = \gamma \hbar B. \tag{1.4}$$

From De Broglie's wave equation, the frequency associated with this energy is

$$\Delta\epsilon = \hbar\omega. \tag{1.5}$$

Therefore,

$$\hbar\omega = \gamma\hbar B. \tag{1.6}$$

Using the subscript zero to indicate the Larmor frequency (ω_0) and the applied external field (B_0) we have

$$\omega_0 = \gamma B_0. \tag{1.7}$$

In any sample of material containing many atoms, the magnetic dipoles will be randomly orientated. When placed in a magnetic field, the dipoles align with the field. For hydrogen nuclei, there are two possible orientations: parallel to the field or in the opposite direction (anti-parallel). Those parallel to the field are in the low energy state and have a small population excess (about ten per two million at 1.5 T) compared to the high energy state (Figure 1.2). For any nuclei in the sample, the magnetic moments of individual nuclei will produce an overall net effect, and it is the behavior of this net effect—the bulk magnetization vector *M*—that is generally considered rather than the behavior of the individual nuclei. Therefore, in the absence of a static magnetic field, *M* will be zero, while in the presence of a field, *M* will have a finite value, pointing in the direction of the field. For convenience this direction is labeled the z-axis. At thermal equilibrium, the distribution of nuclei in the two energy states is given by the Boltzmann distribution

$$\frac{N_{up}}{N_{down}} = e^{-\left(\frac{\Delta\epsilon}{\kappa_B T}\right)}, \tag{1.8}$$

where $\Delta\epsilon$ = energy difference (J), κ_B = Boltzmann constant 1.38×10^{-23}, and T = absolute temperature (K). Therefore, at 23°C (296.15K) and 1.5 T ($\omega_0 = 63.86 \times 106$ Hz) we have

$$\frac{N_{up}}{N_{down}} = e^{-\left(\frac{4.23x10^{-26}}{1.38x10^{-23} x\ 296.15}\right)} = 1.000010345. \tag{1.9}$$

Therefore, for every 1,000,000 in the lower energy state there are approximately 1,000,010 in the upper state. i.e., an excess of 10 per two million (5 ppm) at 1.5 T. A nucleus with spin also resembles a gyroscope in other ways. If tilted from the vertical, a gyroscope will rotate about the vertical axis with a constant angle of tilt. This rotation about the axis is known as precession. In NMR, the individual nuclei that make up *M* make an angle of 54.7° with the z-axis (Figure 1.1). The nuclei therefore precess about z-axis with a characteristic frequency, known as the Larmor frequency. The Larmor frequency is the product of the gyromagnetic constant of the nucleus and the static magnetic field strength. For protons the gyromagnetic ratio is 42.57 MHzT⁻¹, and therefore the Larmor frequency at 1.5 T is approximately 64 MHz.

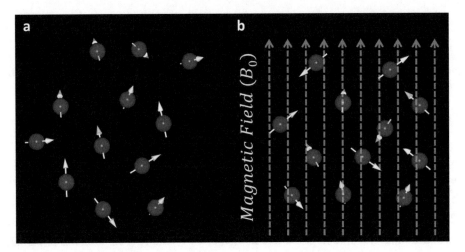

Figure 1.2: Orientation of nuclei before (a) and after (b) being placed in a static magnetic field B0.

1.3   RADIOFREQUENCY NUTATION

In order to detect M in the presence of a strong static magnetic field, M needs to be reoriented perpendicular to the static field, i.e., tipped into the transverse (x-y plane). This tipping can be achieved by applying an alternating magnetic field B_1, at the Larmor frequency, orthogonally, e.g., in the x-direction, to the direction of the static magnetic field B_0 (z-direction). These pulses are referred to as RF pulses. The B_1 field is usually created within a resonant RF transmit "coil" or "antenna," i.e., a tuned LC circuit, that surrounds the object of interest. An electrical current, alternating at the Larmor frequency, is applied to the coil creating a uniform alternating magnetic field within the coil. This resonance condition, i.e., the frequency of B_1 matching the natural Larmor frequency, results in M spiraling down from being aligned along the z-direction into the transverse plane (Figure 1.3a). We often consider a frame of reference rotating at the Larmor frequency, in which case we would "see" M simply being nutated about the axis along which B_1 is applied (Figure 1.3b). The amplitude of an RF pulse (B_1), of duration τ, necessary to cause a rotation of α radians is given by:

$$B_1 = \frac{2\pi\alpha}{\gamma\tau}. \tag{1.10}$$

When an RF pulse is used to initially tip the net magnetization away from B_0 it is usually termed an excitation pulse. However, RF pulses may also be used to invert or refocus magnetization.

The RF transmitter coil is generally large in comparison to the sample in order to create a relatively uniform magnetic field, i.e., excitation flip angle.

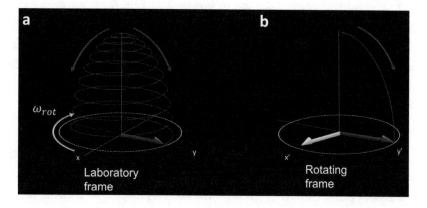

Figure 1.3: (a) The net magnetization **M** (red arrow) is shown as being nutated by 90° by the application of an alternating magnetic field (B_1) (yellow). Note that **M** spirals down into the transverse plane due to the application B_1 rotating at ω_{rot}. In a frame rotating at ω_{rot} the B_1 field appears static and **M** will simply appear as a rotation about the axis that B_1 is applied.

After being excited, a system of spins will return to thermal equilibrium. In 1946, Felix Bloch modeled the magnetization signal in NMR with two decay constants, which he labeled T_1 and T_2 [1-1].

1.4   T$_1$ RELAXATION

T_1 relaxation describes the recovery of the longitudinal magnetization (M_z) back to thermal equilibrium following a perturbation by a RF pulse. Hence, T_1 relaxation is also known as longitudinal relaxation. Furthermore, since T_1 relaxation involves the loss of the energy that was put into the spin system by the RF pulse it is also referred to as spin-lattice relaxation, where the "lattice" in this context consists of surrounding macromolecules. This loss of energy is stimulated by the fluctuating magnetic fields associated with dipole-dipole interactions of neighbouring magnetic moments. T_1 relaxation can only occur when these magnetic field fluctuations occur at the Larmor frequency (ω_0) so T_1 relaxation is very dependent upon the molecular tumbling rate of these dipoles. For example, the protons in water tumble at a wide range of frequencies and there will only be a relatively small number tumbling at ω_0 at any instant of time, therefore the T_1 relaxation time of pure water is quite long. Conversely, the tumbling rate of protons within fat, a much larger molecule, is much slower and therefore more protons tumble nearer ω_0 and hence the T_1 relaxation is more efficient and hence fat has a relatively shorter T_1 relaxation time. Similarly, water molecules that reversibly bind to macromolecules tumble more slowly and hence T_1 relaxation is very inefficient and T_1 relaxation times become longer. The T_1 relaxation time can therefore inform us about the molecular environment of the hydrogen nuclei. The use of exogenous paramagnetic contrast agents, such as chelates of gadolinium, can also be used to improve image contrast. Gadolinium has seven unpaired electrons in its outer shell giving rise to a very large electron magnetic moment. This magnetic moment causes a substantial shortening of the T_1 relaxation time of any water molecules that come into the vicinity of the gadolinium ions.

1.5 T_2 RELAXATION

T_2 relaxation describes the decay of the transverse magnetization (M_{xy}) following an RF excitation. Hence, T_2 relaxation is also known as transverse relaxation. T_2 relaxation does not involve any loss of energy but a loss of phase coherence between the individual spins so that the detectable signal decays with time. Therefore, T_2 relaxation is also sometimes referred to as spin-spin relaxation. As described above, molecular tumbling results in a randomly varying background field that is responsible for T_1 relaxation. Therefore, T_1 relaxation will also diminish the transverse magnetization; hence a component of the T_2 transverse decay can be attributed to T_1 relaxation processes. In addition, there is a secular contribution to T_2. Here the static, or DC, component of the background field starts to dominate and the T_2 relaxation time decreases as the molecular tumbling slows, in distinction to T_1 which starts to increase as the tumbling rate slows.

1.6 T_2* RELAXATION

Once M has been nutated into the transverse plane the individual spins comprising M will dephase with a time constant called T_2^* that includes components due to intrinsic spin-spin interactions (T_2) as well as magnetic susceptibility sources and non-uniformities in the static magnetic field (T_2'). The T_2^* relaxation rate $(\frac{1}{T_2^*})$ can be expressed as the sum of these two relaxation rates

$$\frac{1}{T_2^*} = \frac{1}{T_2} + \frac{1}{T_2'} = \frac{1}{T_2} + \gamma \cdot \Delta B_i \, , \tag{1.11}$$

where ΔB_i represents static magnetic field nonuniformities from all sources. We will discuss T_2^* in more detail later but to simplify the continuing discussion we will only consider pure T_2 relaxation. As discussed in Section 1.6, the transverse decay and longitudinal recovery of magnetization are generally considered as first-order processes with characteristic time constants of T_2 and T_1 respectively. The typical changes in transverse and longitudinal magnetization due to T_2 and T_1 relaxation are shown in Figure 1.4. Note that in biological tissues T_2 relaxation is approximately an order of magnitude shorter that T_1 relaxation.

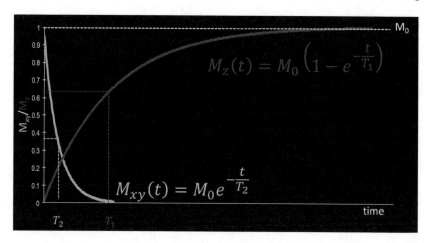

Figure 1.4: T_1 and T_2 relaxation of the longitudinal and transverse magnetization, respectively. The T_1 relaxation time is the time for the magnetization to recover to 69.3% of the equilibrium (M_0) value, while T_2 is the time for the transverse magnetization to decay to 30.7% of M_0.

1.7   BLOCH EQUATION

Bloch derived a differential equation that describes the changes in magnetization during excitation and recovery. The derivation of the Bloch equation is as follows:

The angular momentum ($\boldsymbol{p}$) is changed in time (t) by a torque ($\boldsymbol{T}$)

$$\boldsymbol{T} = \frac{d\boldsymbol{p}}{dt}. \tag{1.12}$$

In the case of a nuclear spin, the spin angular momentum is

$$\boldsymbol{p} = \hbar \boldsymbol{m_I}. \tag{1.13}$$

The torque is expressed via the interaction of the spins' magnetic moment and the external field $\boldsymbol{B}$

$$\boldsymbol{T} = \boldsymbol{\mu} \times \boldsymbol{B}. \tag{1.14}$$

Therefore,

$$\frac{d\hbar \, \boldsymbol{m_I}}{dt} = \boldsymbol{\mu} \times \boldsymbol{B}. \tag{1.15}$$

Since $\boldsymbol{\mu} = \gamma \boldsymbol{p} = \gamma \hbar \boldsymbol{m_I}$,

$$\frac{d\boldsymbol{\mu}}{dt} = \gamma \boldsymbol{\mu} \times \boldsymbol{B}. \tag{1.16}$$

Summing over all spins,

$$\frac{d\boldsymbol{M}}{dt} = \gamma \boldsymbol{M} \times \boldsymbol{B}.$$ (1.17)

Expanding the vector cross-product

$$\frac{dM_x}{dt} = \gamma \left(M_y B_z - M_z B_y \right),$$ (1.18)

$$\frac{dM_y}{dt} = \gamma \left(M_z B_x - M_x B_z \right),$$ (1.19)

$$\frac{dM_z}{dt} = \gamma \left(M_x B_y - M_y B_x \right).$$ (1.20)

We assume that $B_z = B_0$, i.e., the static magnetic field and B_x and B_y are the components of a circularly polarized oscillating magnetic field $B_1(t)$, expressed by

$$B_x = B_1 \cos \left(\omega t \right),$$ (1.21)

$$B_y = -B_1 \sin \left(\omega t \right),$$ (1.22)

$$B_z = B_0.$$ (1.23)

These equations may then be combined to give:

$$\frac{dM_x}{dt} = \gamma \left(M_y B_0 + M_z B_1 \sin \left(\omega t \right) \right),$$ (1.24)

$$\frac{dM_y}{dt} = \gamma \left(M_z B_1 \cos \left(\omega t \right) - M_x B_0 \right),$$ (1.25)

$$\frac{dM_z}{dt} = \gamma \left(-M_x B_1 \sin \left(\omega t \right) - M_y B_1 \cos \left(\omega t \right) \right).$$ (1.26)

These equations are not yet complete since they do not account for relaxation. Bloch [1-1] assumed that spin-lattice and spin-spin relaxation could be treated as a first-order process with characteristic time constants for the decay of T_1 and T_2, respectively. M_x and M_y decay back to their equilibrium value of zero, while M_z returns to its equilibrium value of M_0. The final Bloch equation separated into three componenets is given by

$$\frac{dM_x}{dt} = \gamma \left(M_y B_0 + M_z B_1 \sin \left(\omega t \right) \right) - \frac{M_x}{T_2},$$ (1.27)

$$\frac{dM_y}{dt} = \gamma\left(M_z B_1 \cos(\omega t) - M_x B_0\right) - \frac{M_y}{T_2}, \tag{1.28}$$

$$\frac{dM_z}{dt} = \gamma\left(-M_x B_1 \sin(\omega t) - M_y B_1 \cos(\omega t)\right) - \frac{M_z - M_0}{T_1}. \tag{1.29}$$

As discussed in Sections 1.4 and 1.5, T_1 and T_2 are often called the longitudinal and transverse relaxation times, respectively, since they are the time constants for decay of the components of magnetization along and perpendicular to B_0. The Bloch equations can be solved by straightforward but laborious procedures under various limiting conditions. A simple case to consider is if we nutate or tip the spins through 90° using a short duration radio-frequency pulse so that there is negligible relaxation during the pulse. In this case the equations simplify to

$$\frac{dM_x}{dt} = \omega M_y - \frac{M_x}{T_2}, \tag{1.30}$$

$$\frac{dM_y}{dt} = -\omega M_x - \frac{M_y}{T_2}, \tag{1.31}$$

$$\frac{dM_z}{dt} = -\frac{M_z - M_0}{T_1}. \tag{1.32}$$

The solutions to these differential equations are

$$M_x(t) = M_0 \sin(\omega t) e^{-\frac{t}{T_2}}, \tag{1.33}$$

$$M_y(t) = M_0 \cos(\omega t) e^{-\frac{t}{T_2}}, \tag{1.34}$$

$$M_z(t) = M_0\left[1 - e^{-\frac{t}{T_1}}\right]. \tag{1.35}$$

The longitudinal magnetization $M_z(t)$ returns exponentially to its equilibrium value M_0, while the transverse magnetization $M_x(t)$ and $M_y(t)$ behaves as an exponentially damped sinusoidal oscillation decaying to its equilibrium value $M_{xy} = 0$. In complex notation,

$$M_{xy} = M_x + iM_y, \tag{1.36}$$

$$M_{xy}(t) = M_0 e^{i\omega t} e^{-\frac{t}{T_2}}. \tag{1.37}$$

1.8   SIGNAL DETECTION

The precessing transverse magnetization induces a small alternating voltage in a receiver coil in accordance with Faraday's Law of electromagnetic induction. The magnitude of the transverse magnetization will decrease with a time constant given by T_2^*, as discussed previously, resulting in a decaying signal known as a free induction decay (FID). This FID signal is then amplified, digitized, and processed to extract the required frequency, phase, and amplitude information for the desired application. The receiver coil may be the same coil as used for RF transmission or it may be a separate array of coils, that more closely fits the sample.

In this section, we cover topics that are relevant to each of the specific hardware components, such the static superconducting magnets, the magnetic field gradients and parameters that limit them, radiofrequency coils and signal-to-noise.

1.8.1   MRI HARDWARE

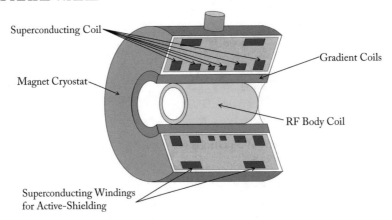

Figure 1.5: A schematic diagram showing a cross-section through a superconducting MRI system. B_0 is created by a current flowing through a set of superconducting coil windings. The fringe field of the magnet is partially mitigated by a second set of superconducting coil windings with current flowing in the opposite direction. Both sets of coils are inside a liquid helium-filled cryostat. Concentric cylinders carrying the three main gradient coil sets are mounted inside the cryostat. The main "body" RF coil of birdcage design is mounted inside the gradient coils, closest to the subject being imaged.

An MRI system comprises four main components. The first is an appropriate magnet to generate the strong static magnetic field (B_0), which is required to induce the nuclear polarization. The second is a RF system that generates the required alternating magnetic field (B_1), at the Larmor frequency, and detects the weak MR signal being returned from the patient. The third component is the gradient system that generates the required linear magnetic field variations that are superimposed upon B_0 and are used to spatially encode the MR signal. The gradient and whole-body RF coils are typically concentrically positioned inside the bore of the magnet. Figure 1.5 shows a schematic cross section through a superconducting MRI system. Finally, the fourth component consists of several computers that are used to provide the user interface, generate the digital representations of the RF and gradient pulses, and perform the mathematical operations, e.g., Fourier transformation, required to reconstruct an image from the digitized signals returned from the patient.

1.8.2   SUPERCONDUCTING MAGNETS

The requirements for the static field are that it should be reasonably strong, stable, and uniform. Magnetic field strength or, more accurately, "magnetic flux density," is measured in the SI-derived unit of tesla (symbol T). While it is certainly true that bigger is not necessarily better, there is an approximate linear increase in signal-to-noise ratio (*SNR*) with increasing field strength. Over the last 30 years of MRI development, various technologies and designs have been used to create magnets of an appropriate configuration that a human body can be placed inside. These designs have various trade-offs in terms of field strength as well as stability, uniformity, and patient acceptance. It was not until the development of superconducting magnet technology that higher field strengths could be achieved. Typical clinical superconducting MRI systems are available at 1.5 T and 3.0 T, having Larmor frequencies of 64 and 128 MHz, respectively. Most vendors have also developed 7.0 T whole body systems for research purposes, with even higher field strengths, e.g., 11.7 T being custom-developed. Superconducting magnets generate their magnetic field by circulating an electric current through solenoidal coils of niobium-titanium (NbTi) filaments embedded in a copper matrix. At temperatures below approximately 10 K NbTi becomes superconducting, which means it has zero electric resistance and current will circulate indefinitely without an external power supply. Practically, a superconducting MRI magnet comprises a steel cryostat with the superconducting coils immersed in liquid helium at 4.2 K (-269°C or -452°F). While iron-cored or permanent magnet designs can have a C- or H-shape and may therefore appear more patient friendly, superconducting MRI system cryostats are most commonly shaped like a cylindrical tube with the patient placed inside the central bore. In addition to the patient, the gradient and RF body coil are also positioned inside the bore, although hidden from the patient behind the system covers. Although the diameter of the cryostat bore is approximately 100 cm, once these additional coils are installed the patient accessible bore is reduced to approximately 60 cm. However, in response to demands for improved patient acceptance as well as an increasingly bariatric population, manufacturers have introduced systems to allow 70 cm or greater patient apertures as well as reducing overall magnet lengths.

While a strong, but not necessarily very uniform, B_0 is required to achieve a good nuclear polarization, an extremely high uniformity is required to perform spatial localization. The uniformity of B_0 is usually defined as the variation of the field within a spherical volume of a given diameter which, for a 1.5 T magnet would typically be less than 1 part per million (ppm) over a 40 cm diameter spherical volume. Further improvements can be achieved over smaller volumes by a process known as "shimming," which is often automatically performed by the system for each body part imaged. In addition, the field must be temporally very stable, with a typical superconducting magnet having a stability of better than 0.1 ppm/h.

While the signal-to-noise ratio increases with field strength, often allowing higher spatial resolution, higher magnetic fields, e.g., 3 T and above, also bring additional challenges. These include: a lengthening in T_1 relaxation times resulting in poorer T_1-weighted image contrast; increased magnetic susceptibility effects that can cause regional signal loss and localized geometric nonlinearity; increased artifacts associated with involuntary motion, physiological flow, and variation in the uniformity of RF excitation.

Magnets also have an associated fringe-field that needs to be considered when siting the system. Most superconducting magnets are now actively shielded which means they have a separate set of superconducting coils inside the cryostat, but positioned outside the main magnet coils, in which the current flows in the opposite direction, reducing the magnitude and effect of the external magnetic field on the surrounding environment. However, due to siting space limitations, occasionally additional passive magnetic shielding in the scan room wall(s) may be required.

The MR system must be situated within a six-sided RF-shielded examination room with conductive metal lining, usually made of copper or aluminium through which external RF electromagnetic interference will not pass.

1.8.3   MAGNETIC FIELD GRADIENTS

Magnetic field gradients are used for altering the Larmor frequency over the sample, which allows spatial localization, as discussed in more detail in Chapter 2. Linear magnetic field gradients are created by additional coils of wire which are positioned inside the magnet bore, adjacent to the liquid helium cryostat in a cylindrical superconducting system. These coils are designed to provide gradients in the three orthogonal physical directions x, y, and z. The effective fulcrum point of each of the three gradients is at the center of the magnet bore, known as the isocenter. Gradient pulses require time for the gradient field to ramp up to the desired amplitude and then ramp down again, hence a very common gradient pulse design is a trapezoid. Gradient performance is defined by the maximum achievable amplitude of the gradient in mTm^{-1}; typically in the range 30–50 mTm^{-1} and the time taken to ramp up and down the waveform, known as the rise-time, typically in the range 200–100 µs. Higher gradient amplitudes allow the system to achieve thinner slices or smaller fields-of-view, while shorter rise-times allow faster encoding. Depending upon the desired imaging plane the required logical gradient pulses: slice-selection, phase, and frequency encoding, are played out on the appropriate physical x, y, and z gradients, with a mixed combination for oblique imaging.

Gradient switching is accountable for the characteristic knocking noise heard during MRI. As the coils lie within strong static magnetic fields and currents in the coils are pulsed, the coil windings experience a Lorentz force which causes mechanical vibrations, albeit of very small amplitude, of the gradient coils resulting in acoustic noise. The gradient switching also induces currents in the magnet cryostat, known as eddy currents. These currents decay with time but also cause their own magnetic fields, which can cause image artifacts. Gradient coils are themselves actively shielded by a second set of outer coils to minimize the induction of unwanted eddy currents in the cryostat and other conducting structures.

Gradient coils require high currents and voltages in order to produce field variations over the size of the human body. Hence, amplifiers are needed to convert the digital waveforms into gradient pulses. Most of this power is dissipated as heat and the gradient coils and often the amplifiers require water-cooling, and the maximum allowed peak power is limited by equipment and patient safety requirements. The maximum duty cycle (DC) is another limitation used to prevent excessive heating of the gradient coils and amplifiers from damaging the hardware, which is usually defined as the root-mean-square of the gradient patterns over a time period:

$$DC = \frac{\int_{t_1}^{t_2} G(t)^2 \, dt}{G_{max}^2 \left(t_2 - t_1\right)} \cdot 100\%.$$

(1.38)

1.8.4   RADIOFREQUENCY (RF) COILS

The RF transmitter creates shaped pulses with specific amplitude and phase, centered at the Larmor frequency, which can be used to tip the net magnetization. The RF pulse waveform is amplified and applied to the transmitter coil. The transmitted RF magnetic field is often referred to as B_1^+. Most commonly this is a large diameter coil, known as the body coil, which is located just inside the gradient coil assembly. While the MR signal received back from the patient can be detected by the body coil, its large size means that the received SNR would be relatively poor. In order to maximize SNR, the receiver coil is often positioned closely to the anatomy of interest, hence the large number of anatomically optimized receiver coils available (e.g., head, spine, shoulder, knee, etc). In recent years, most receiver coils have been constructed from arrays of smaller coil elements. The idea is that smaller coils have a better SNR but a limited field-of view (FOV). Careful combination of multiple coils in a matrix configuration gains the advantage of small coil SNR but with a larger FOV. The complex, i.e., the real and imaginary, signals from each individual coil element are amplified, digitized and then reconstructed. To minimize noise induced in the receiver chain, detected signal digitization occurs near the RF receiver coil(s). Although the signal is centered on the Larmor frequency, the MRI encoding process only involves a small range of frequencies typically in the range ±16 kHz to ±250 kHz, known as the receiver bandwidth. Part of the receiver processing is to extract this small readout bandwidth before passing the digitized signals from each coil to the reconstruction hardware. Typically, the individual images from each coil are combined using a root sum-of-squares algorithm [1-3]. However, individual coils also have different spatial sensitivity profiles. The receive coil profiles are often referred to as B_1^-. This differential sensitivity can then be utilized in various methods employed to accelerate the MR acquisition. This is discussed in more detail in Chapter 4.

1.8.5   SIGNAL-TO-NOISE (SNR)

There is a limited *SNR* that limits the achievable resolutions within reasonable imaging times. *SNR* is defined as the ratio of the signal to the standard deviation of the noise:

$$SNR = \frac{S_{signal}}{\sigma_{noise}}. \tag{1.39}$$

The signal is directly proportional to the total number of polarized spins, or net magnetization:

$$S_{signal} \propto M_0. \tag{1.40}$$

Random thermal fluctuations in the sample and electronics are a major source of noise in MRI [1-3, 1-4]. Electronic noise is dominated by Brownian motion of electrons, causing Johnson noise described by

$$\sigma_{noise}^2 = 4kRT\Delta f, \tag{1.41}$$

where k is Boltzmann's constant, T is the temperature in Kelvins, R is the resistance of the sample or electronics, and Δf is the receiver bandwidth.

Combining these, *SNR* is therefore proportional to the total magnetization, and inversely by the square-root of temperature and receiver bandwidth:

$$SNR \propto \frac{M_0}{\sqrt{4kRT\Delta f}}. \tag{1.42}$$

Assuming a cubic sample, increasing the size of the sample dimensions increases the net magnetization that is measured, which results in a directly proportional increase in the overall signal:

$$SNR \propto \frac{M_0}{\sqrt{4kRT\Delta f}} \Delta x \Delta y \Delta z \tag{1.43}$$

When repeated acquisitions have independent noise distributions, the signal increases with the number of acquisitions. The noise also increases, although in quadrature, which becomes the square-root of the number of repeated acquisitions:

$$SNR \propto \frac{M_0}{\sqrt{4kRT\Delta f}} \sqrt{N_{acq}}. \tag{1.44}$$

It is useful to describe *SNR efficiency*, which is *SNR* multiplied by the square root of the total acquisition time:

$$SNR = \frac{1}{\sqrt{T_{acq}}} \frac{S_{signal}}{\sigma_{noise}}. \tag{1.45}$$

With temporal acquisitions, a standard *SNR* calculation is a challenge because there are many different types of contrast weightings possible, as discussed in Chapter 3. Temporal *SNR* (*tSNR*) is one method to describe changes in time, which is used within functional MRI communities [1-5]. When comparing similar acquisition schemes with a single variation, we can compare *tSNR*, which is defined as:

$$tSNR = \frac{\overline{S}_{signal}}{\sigma_{noise}}. \tag{1.46}$$

SNR is important to recognise as one fundamental limitation in MRI, as it limits the ability to encode signals quickly while still obtaining sufficient, useful information regarding the patient.

1.9   CONCLUSION

In this chapter, we have described the basic principles of a nuclear magnetic resonance and the experimental setup. In the following chapter, we will investigate the use of gradient fields to encode spatial locations within a sample in order to obtain magnetic resonance images.

BIBLIOGRAPHY

[1-1]   F. Bloch, Nuclear induction, *Phys. Rev.*, 70(7–8), pp. 460–474, Oct. 1946. DOI: 10.1103/ PhysRev.70.460. 1, 5, 8

[1-2]   E. M. Purcell, H. C. Torrey, and R. V. Pound, Resonance absorption by nuclear magnetic moments in a solid, *Phys. Rev.*, 69(1–2), pp. 37–38, Jan. 1946. DOI: 10.1103/Phys-Rev.69.37. 1

[1-3]   P. B. Roemer, W. A. Edelstein, C. E. Hayes, S. P. Souza, and O. M. Mueller, The NMR phased array, *Magn. Reson. Med.*, 16(2), pp. 192–225, Nov. 1990. DOI: 10.1002/ mrm.1910160203. 14

[1-4]   D. Hoult and R. Richards, The signal-to-noise ratio of the nuclear magnetic reso-nance experiment, *J. Magn. Reson.*, 24(1), pp. 71–85, Oct. 1976. DOI: 10.1016/0022-2364(76)90233-X. 14

[1-5]   M. Welvaert and Y. Rosseel, On the definition of signal-to-noise ratio and contrast-to-noise ratio for fMRI data, *PLoS One*, 8(11), p. e77089, Nov. 2013. DOI: 10.1371/journal.pone.0077089. 15

CHAPTER 2

Spatial Encoding

In Chapter 1, we discussed the fundamentals of NMR. In this chapter, we discuss how to obtain data for the formation of images with spatial encoding performed using magnetic field gradients.

2.1 INTRODUCTION

Magnetic resonance imaging requires the spatial localization of the NMR signal. Both data acquisition and reconstruction are intimately tied together through the spatial encoding method. A pulse sequence is a combination of RF and gradient pulses. Generally, the amplitudes and timings between RF pulses will dictate the image contrast whereas the gradients will primarily be responsible for spatially encoding the signals. This is a simplification since gradients can also dephase/rephase transverse magnetization and hence influence image contrast. This will be discussed in Chapter 3, while here we focus on the various ways in which the signal can be spatially encoded.

Since the frequencies of NMR signals depend on the local magnetic field, both Lauterbur [2-1] and Mansfield and Grannell [2-2] proposed the use of a linear magnetic field gradient to localize the signal. Indeed, this relationship between spatial position and frequency led Lauterbur to use the term "*zeugmatography*" [2-3] from the Greek ζεῦγμα "*that which is used for joining.*" Lauterbur obtained two-dimensional (2D) images by rotating the gradient relative to the object of interest thereby obtaining a series of projections, Fourier transforming each and then reconstructing the image by back-projection. Although this method produced a 2D image there was no spatial localization in the third direction. This required the development of selective excitation of a slice through a sample and was first proposed by Lauterbur et al. [2-4] and Mansfield et al. [2-5]. Although the method of back-projection is still in use (see Section 2.3.5) the standard method of acquiring 2D images is based upon slice-selective 2D Fourier imaging, first proposed by Kumar et al. [2-6] and subsequently refined by Edelstein et al. [2-7] into the "spin-warp" method.

2.2 CARTESIAN ENCODING

The basic mathematical theory behind 2D Fourier imaging is derived below. Consider a magnetic field gradient, along the x-direction. Note that although the gradient is along x, the magnetic field lies along the same direction (z) as the static magnetic field B_0, i.e.,

$$\frac{\partial B_z}{\partial x} = G_x.$$

(2.1)

When this gradient is superimposed on the static magnetic field B_0, the magnetic field at a location x given by $B(x)$, can be expressed by

$$B(x) = B_0 + G_x x .$$ (2.2)

The spins at position x will resonate at a frequency given by

$$\omega(x) = \gamma \left[B_0 + G_x x \right],$$ (2.3)

or in a reference frame rotating at frequency ω_0, the precessional frequency at location x becomes

$$\omega(x) = \gamma G_x x .$$ (2.4)

Let the spin density at location x be $\rho(x)$, in which case the signal for spins between location x to $x + \delta x$ can be written in complex form as

$$dS(x) = \rho(x) e^{-i\omega(x)t} dx = \rho(x) e^{-i\gamma G_x xt} dx ,$$ (2.5)

or for the entire object from x to $x + \delta x$

$$S(x) = \int_x^{x+\delta x} \rho(x) e^{-i\gamma G_x xt} dx .$$ (2.6)

These equations show that $S(x)$ is the Fourier Transform of $\rho(x)$. In the general case of a continuous distribution of spin density, the signal with 2D orthogonal encoding becomes

$$S(x,y) = \iint \rho(x,y) e^{-i\gamma G_x xt} e^{-i\gamma G_y yt} dxdy.$$ (2.7)

The spin density, $\rho(x)$, incorporates all effects within the spin density, such as T_1 and T_2 relaxation.

By applying the gradients over a period of time, a spatially dependent phase shift can be created, which is used to encode the spins in different locations of the sample.

$$\Delta\phi(x) = \gamma \int_{t_1}^{t_2} G_x(t) \cdot x \ dt .$$ (2.8)

This phase shift can be applied either immediately before or during an acquisition. As a convention, it is helpful to consider this phase shift in a domain called k-space.

2.2.1   *k*-SPACE: SPATIAL FREQUENCY DOMAIN

The signal equation in (2.7) is often written to draw a more direct relationship with the Fourier transformation, which transforms between spatial locations, x and y, and spatial frequency parameters, k_x and k_y:

$$S(x,y)=\iint\rho(x,y)e^{-ik_x x}e^{-ik_y y}dxdy. \tag{2.9}$$

The spatial frequencies k_x and k_y are thus defined as follows:

$$k_x=\gamma\int G_x(t)dt\,, \tag{2.10}$$

$$k_y=\gamma\int G_y(t)dt\,. \tag{2.11}$$

The definition of k_x and k_y can often vary by a factor of 2π, where some definitions use $\bar{\gamma}$ instead of γ in Equations (2.9) and (2.10), although Equation (2.7) would remain the same. The definition above is used to keep derivations more compact.

In the simplest case of gradients with constant amplitude these equations simplify to

$$k_x=\gamma G_x t\,, \tag{2.12}$$

$$k_y=\gamma G_y t\,. \tag{2.13}$$

Thus, the spatial frequencies are proportional to the gradient strength and duration to which the spins are subjected. The units of k-space are radians cm^{-1}, therefore an alternative term to spatial frequency would be spatial phase gradient, since k_x and k_y represent the phase advance or retardation that the spins experience per cm of object in the x and y directions. We can now re-write the 2D spatial encoded signal as

$$S(k_x,k_y)=\iint\rho(x,y)e^{-ik_x x}e^{-ik_y y}dxdy. \tag{2.14}$$

This is defined for both Cartesian and non-Cartesian encoding methods.

Many pulse sequences acquire data with regular $k(t)$ sampling such that it encodes with a grid-like or Cartesian sampling pattern. Cartesian encoding enables predictable gradient waveforms that can use the very efficient reconstruction method of the Discrete Fourier Transform (DFT; see Chapter 4). Figure 2.1 shows a Cartesian k-space acquisition scheme, and several non-Cartesian acquisition trajectories.

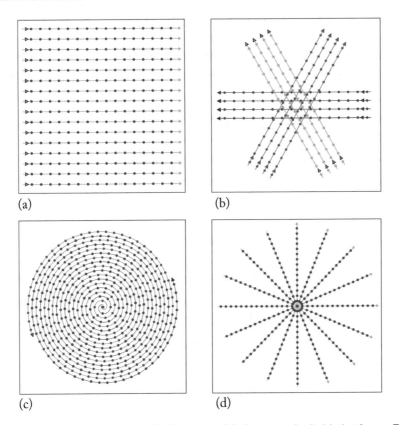

(a)   (b)

(c)   (d)

Figure 2.1: Different *k*-space trajectories: (a) Cartesian; (b) Cartesian/radial hybrid, e.g., PROPEL-LER (GE) or BLADE (Siemens) or VANE (Philips); (c) spiral with two constant angular velocity interleafs (red and blue); and (d) projection reconstruction (radial) showing eight equally spaced acquisitions.

All the spins within a sample can be excited with a "non-selective" excitation, which usually involves a rectangular (or "hard") RF pulse. However, to reduce the amount of encoding needed, a single 2D slice or three-dimensional (3D) slab is usually excited before encoding the sample in the spatial frequency domain. RF pulses are usually combined with gradient pulses in order to select a slice of the sample (Figure 2.2). While applying a gradient along an axis, the excitation RF pulse is usually modulated with the Fourier transform of the desired slice profile. The amplitude of the gradient and the bandwidth of the pulse determines the thickness of the excited slice,

$$\Delta z = \frac{\Delta \omega_{RF}}{\gamma G}.$$ (2.15)

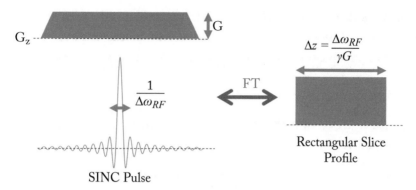

Figure 2.2: A slice selection gradient and SINC RF pulse are shown, which causes a rectangular spatial region, Δz, to be excited.

2.2.2   SPIN WARP IMAGING

Spin warp imaging is a fundamental MRI technique to understand for Cartesian encoding. The basic principle of a gradient echo-based slice-selective spin warp imaging sequence is shown in Figure 2.3. For a gradient echo (*GRE*), spatially dependent phase shifts are created by the application of a gradient in each spatially-encoded direction. In order to acquire both positive and negative k-space, the frequency (nominally "x") encoded direction is dephased, and then rephased when sampling a single line of k-space. The time between the excitation RF pulse and the signal rephasing is called the "echo time" or "*TE*." The process of acquiring a single line of k-space is repeated over distinctive lines in the phase (nominally "y") encoded direction to obtain the entire region before image reconstruction. The time between subsequent slice/slab selective RF pulses is referred to as the "repetition time" or *TR*.

In order to obtain an image under full Cartesian sampling, the *FOV* is proportional to the interval between k-space points. The Nyquist–Shannon theorem states that a band-limited signal with bandwidth, B, can be completely reconstructed from its samples if they are sampled at a rate no greater than $1/2B$, hence defining the achievable image resolution for a fully-sampled experiment of discrete points. Because sampling of k-space is discrete, the number of samples is dictated by the desired number of pixels while satisfying the Nyquist theorem (see Chapter 4). For the frequency encoding (*FE*) direction, the amplitude of the frequency encoded gradient is given by

$$G_{FE} = \frac{1}{\gamma FOV_{FE} \Delta t},$$ (2.16)

where FOV_{FE} is the field of view in the frequency encoding direction and Δt is the total sampling time. Since only one phase encoding is performed per *TR*, in order to fully cover k-space the phase encoding gradient needs to be performed N_{PE} times, where N_{PE} is the total number of phase encoding steps. For each phase encoding step the area of the gradient needs to be incremented. The maximum amplitude of a phase encoding gradient pulse of duration τ is given by

$$G_{PE} = \frac{1}{\gamma FOV_{PE}\tau},$$ (2.17)

where FOV_{PE} is the field of view in the phase encoding direction. The step amplitude of each phase encoding gradient is given by

$$G_{PE,n} = n\frac{G_{PE,max}}{N_{PE}} \quad \text{for} \quad -\frac{N_{PE}}{2} \leq n \leq \frac{N_{PE}}{2}.$$ (2.18)

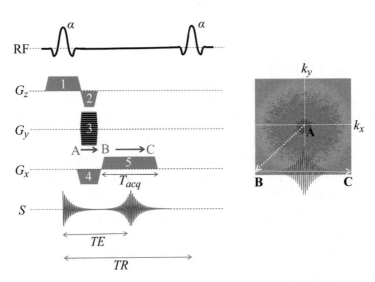

Figure 2.3: Gradient echo slice selective spin warp imaging sequence and associated k-space trajectory. The initial excitation pulse in the presence of the G_z slice select gradient (1) nutates the spins into the transverse plane (time point A). The negative polarity slice select rephasing gradient (2) compensates for the phase shift across the slice induced by (1). The G_y phase encoding gradient (3) and the G_x frequency encoding prephasing gradient (4) moves the spins the appropriate distance in k_x and k_y (time point B). In this case the maximum negative phase encoding gradient has been applied. The gradient echo signal is formed in the presence of the G_x frequency encoding gradient (5) during time points B to C. The gradient echo is sampled during the frequency encoding gradient (5). A second excitation pulse repeats this cycle at the repetition time ("TR") to obtain additional k-space lines each with a different amplitude phase encoding gradient.

It is also possible to acquire true 3D images by applying a second phase encoding gradient in the slice select (nominally "z") direction. Like the in-plane phase encoding process, this gradient is applied as a number of steps equal to the desired number of partitions (or slices) within the volume

(N_{SS}). This creates a 3D k-space and the images are reconstructed using a 3D Fourier transform. For example, if we excite a slab of tissue 64-mm thick and apply 32 phase encodings then, following Fourier transformation, we would obtain 32 slices with an effective slice thickness of 2 mm. Since we are exciting a large slab of tissue the *SNR* is improved by $\sqrt{N_{SS}}$ relative to a 2D acquisition of the same effective slice thickness. However, the acquisition time would be proportional to $N_{PE} \cdot N_{SS}$ · *TR*. 3D imaging is primarily used with pulse sequences that have short *TR*s, e.g., gradient echoes due to the large number of phase encoding steps.

Within a single 2D slice the simplest k-space encoding method, described above, is to encode a single line, or echo, in the frequency encoding direction in each *TR* period and to step the phase encode gradient between *TR*s to move between lines. The spin-warp method is relatively slow but is the basis of acquiring most clinical MR images. Other methods exist that can fill k-space faster, but they invariably require trade-offs in terms of image quality.

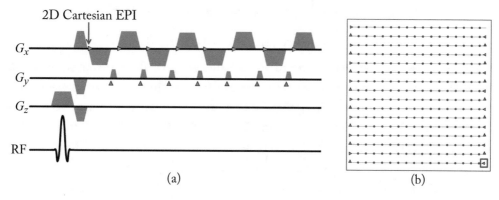

Figure 2.4: A 2D Cartesian EPI pulse sequence (left) and k-space (right). After the initial excitation, the frequency encoding gradient is swept back and forth to cover k-space, while a "blipped" phase encoding gradient increments the k-space after each line is swept through. The red arrow and box indicate the beginning location of the full k-space acquisition, while the green/orange arrows indicate the direction that k-space is traversed.

Echo-planar imaging (EPI), for example, is a fast imaging method that applies gradients "blips" on the phase-encode axis and reversals of the frequency encoding gradient to acquire all the k-space lines of an image following a single excitation (Figure 2.4). However, EPI images demonstrate spatial distortions due to spins that are off-resonance from the nominal Larmor frequency, due for example, to B_0 non-uniformity. The EPI sequence begins very similarly to a gradient-echo spin warp sequence, with conventional slice selection and phase encoding gradient steps. However, unlike spin-warp imaging, k-space is swept back and forth to cover a larger region within a single *TR*, which can include any number of k-space lines. The number of k-space lines that are acquired per *TR* is referred to as the "echo train length (ETL)", "EPI factor", or "shot factor". The fraction of

k-space acquired is referred to as "shots" or "segments." Thus, for a full k-space with 256 lines, where 128 lines are acquired per *TR*, the EPI factor is 128, and the number of shots is 2. EPI has much faster acquisition times than a standard GRE; however, due to its longer readout, EPI is subject to artifacts caused by non-uniform fields at the boundaries of tissues, particularly where air and tissue have large magnetic susceptibility differences.

2.3   NON-CARTESIAN ENCODING

Non-Cartesian acquisitions have several advantages over traditional Cartesian approaches, particularly for:

- improved resilience against motion,

- ultra-short/zero echo time imaging (useful for experiments with extremely short T_2 relaxation time, such as bone, lungs, or other nuclei such as sodium), and

- incoherent sampling for compressed sensing (see Chapter 4).

Non-Cartesian methods can also avoid artifacts due to signal aliasing from structures larger than the *FOV* due to their oversampling of the center of k-space, since the *FOV* increases with the inverse of the k-space density. Acquiring k-space from the center out is also more robust against motion, especially in 3D acquisitions. Since non-Cartesian encodings do not follow the concept of phase/frequency encoding, artifacts tend to be more uniformly and incoherently distributed rather than the coherent artifacts seen along the phase encode direction in spin-warp imaging.

The prominent disadvantage of non-Cartesian acquisitions is that they require more complex system implementations for both acquisition and reconstruction. Intuition developed from Cartesian imaging is often lost, although metrics such as sampling efficiency provide an immediate translation. Some formalisms also differ, e.g., the *TE* can be defined as either the point when sampling first occurs or the time when the center of k-space is sampled, such as multi-echo experiments within a single *TR*.

Within a non-Cartesian 2D encoding there are several k-space trajectories that have been explored including radial [2-1], [2-8–2-10], spiral [2-11–2-13], concentric rings [2-14], Lissajou [2-15], and leaf-like petals [2-16, 2-17], with many others described in the literature. Here we will focus on the more established methods of radial and spiral trajectories, shown in Figure 2.5, to provide an understanding for developing more advanced techniques.

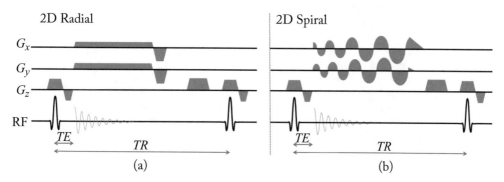

Figure 2.5: A 2D gradient spoiled gradient recalled echo sequence showing (a) radial and (b) spiral acquisitions with a pulse sequence schematic. The *TE* is defined as the time between the middle of the excitation pulse and the beginning of acquisition at the center of k-space. The k-space trajectory is rewound so that the time integral of the x and y gradients are zero at the end of the sequence. Gradient spoiling often follows rewinding in order to reduce any remaining transverse magnetization before the next excitation.

2.3.1   2D RADIAL ENCODING

In his seminal 1973 paper, Lauterbur rotated an object about an axis relative to the gradient field from which data was then used to create the first projection reconstruction (PR) image. PR and radial encoding are two highly related non-Cartesian techniques. PR obtains its name from similarities with CT reconstruction. PR prephases the gradients before acquiring both negative and positive k-space values, while radial traditionally means originating from the center of k-space (Figure 2.6). Therefore, PR acquires half the number of k-space lines as radial sampling but acquires the same number of data points. The choice between the two is dictated by imaging needs: PR acquires more points per excitation, while radial has less T_2^* signal decay. Zero echo time (ZTE) imaging uses a 3-D radial center-out acquisition, ramping the gradient during excitation, thus starting slightly off-center when the acquisition begins.

Radial sampling has advantages over many other methods: it samples the center of k-space before significant T_2^* decay, for example allowing it to capture signal from cortical bone and demonstrating reduced susceptibility induced signal dephasing which is important for lung imaging; it can also be used in a multiple echo mode for quantitative T_2^* mapping. PR has been widely used for motion reduction [2-8], dynamic measurements, and image undersampling [2-1] (see Chapter 4; also see [2-9, 2-10]).

Projection Reconstruction

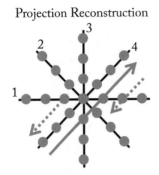

Radial

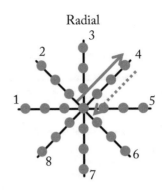

Figure 2.6: Projection reconstruction (PR, left) and radial (right) sampling are shown here with equi-distributed spokes. PR starts sampling on the other side of the center of k-space, while radial conventionally starts sampling at the center of k-space. A simple, uniform sampling pattern evenly divides the possible angles, 180° or 360°, by the total number of spokes and increments the acquired spoke linearly (imaging spoke 1 first, followed by spoke 2, 3, etc.)

The gradient requirements for radial sampling are very generally low, which means the acoustic noise and peripheral nerve stimulation is low. The gradients are simply rotated by an increment of θ to create the individual spokes

$$k_{x,i}\left(t\right)=tk_{max}\sin\theta_{i}\,, \tag{2.19}$$

$$k_{y,i}\left(t\right)=tk_{max}\cos\theta_{i}\,. \tag{2.20}$$

A simple approach for the ordering of θ is to divide 2π by the number of k-space readouts or spokes ($= N$), and then increment each spoke by that factor. This results in even distribution of spokes. The ith ($i = 1\ldots N$) spoke is then acquired at the following angle:

$$\theta_{i}=2\pi\frac{i}{N}. \tag{2.21}$$

Alternative rotation strategies such as the golden angle have also been explored (see below). Consideration must be given to the ramping of the gradient, which effects k-space estimations, especially if different bandwidths, slew rates, or gyromagnetic ratios are considered.

2.3.2   RADIAL FIELD OF VIEW

The Nyquist limit states that the sampling frequency must be twice the frequency being sampled. In Cartesian coordinates, the frequency encoding, or image resolution, must therefore be twice that of the maximum k-space value:

$$\text{Image resolution*2} = 2\Delta x = \frac{1}{k_{max}}. \tag{2.22}$$

The resolution can be estimated from the maximum gradient and the time for each radial spoke:

$$\Delta x = \frac{FOV}{N} = \frac{1}{2} \cdot \frac{\gamma}{2\pi} \cdot G_{max} \cdot T_{sampling}. \tag{2.23}$$

This Cartesian assumption works well as a rule-of-thumb.

Both the *FOV* and resolution are anisotropic, with radial and angular components (Figure 2.7). The area contained within the *FOV* is the area between four adjacent *k*-space points:

$$\Delta S = \Delta k_r \cdot \left(k_r \Delta k_\theta \right) = \frac{1}{FOV} \cdot \frac{1}{FOV}. \tag{2.24}$$

The additional k_r on the angular component accounts for the change in density as more distant *k*-space points are acquired.

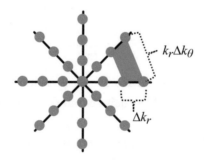

Figure 2.7: A radial *k*-space trajectory consists of volumetric elements that create the Nyquist field-of-view. The *k*-space area changes with the density of the radial lines, Δk_r, and the angular distance between adjacent radial lines, $k_r \Delta k_\theta$, which increases as the distance from the center increases. It is worth noting that radial sampling creates *FOV*s that have a spatial frequency dependency. In other words, the *FOV*s are non-uniform, being larger for low-frequency spatial changes due to their increased *k*-space density, and high-frequency image components such as image edges requiring more samples to avoid image aliasing. The Nyquist *FOV* is calculated based on the highest frequency that we wish to image, and so we base all *FOV* calculations on the *k*-space area at the *k*-space extremities.

A simplification is made in that the spacing between each radial *k*-space component, Δk_r, is usually much higher than the radial *FOV*. This has a drawback where continuous sampling creates a large amount of stored data, which may be important for increasing resolutions and coil channels. Under this assumption, the *FOV* constraints are determined largely by the angular component of *k*-space:

$$k_r \Delta k_\theta = \frac{1}{FOV} \quad . \tag{2.25}$$

It is worth noting that this shows that low-frequency portions of the image do not alias due to the high central sampling densities. Low-frequency components will often alias due to image reconstruction constraints. The *FOV* is still limited to avoid aliasing of high-frequency features, such as boundaries between tissues, which can cause speckling artifact. The difference between two points at the edge of *k*-space creates the angular *FOV*:

$$k_{max}(1 - \Delta k_r)\Delta k_\theta = \frac{1}{FOV} \quad . \tag{2.26}$$

Under a similar assumption that Δk_r approaches zero with maximum sampling, the *FOV* becomes:

$$FOV \leq \frac{1}{k_{max}\Delta k_\theta}. \tag{2.27}$$

The number of spokes can be used to determine the *FOV* limits, because $\Delta k_\theta = \frac{2\pi}{N}$ for equidistant radial sampling:

$$FOV \leq \frac{N}{2\pi k_{max}} \quad . \tag{2.28}$$

Thus, to fulfill the Nyquist limit and avoid aliasing with radial encoding, the rule of thumb is that the number of radial lines should be equivalent to the number of Cartesian lines multiplied by $\pi/2$.

With the assumption of equidistribution and a fixed *FOV*, *SNR* increases with the square of the number of *k*-space lines, but decreases with the square of the reconstructed image matrix:

$$SNR \propto \frac{1}{\sqrt{R}} = \sqrt{N_{spokes}} / \sqrt{N_{image-size}} \quad . \tag{2.29}$$

2.3.3   EFFICIENCY OF 2D RADIAL *k*-SPACE ACQUISITIONS

Radial sampling is less efficient than Cartesian sampling because it requires more excitations/lines to achieve the Nyquist limit.

The area of a circular region is πr^2 that we have normalized so that the maximum $r = 1$. The number of samples required to encode this with 2D PR is πN. 2D radial sampling requires encoding each point twice as frequently as 2D PR, which is $2\pi N$. The density of each radial point decreases with increasing r, as shown in the *FOV* and resolution estimations. We can calculate the density required to cover the area:

$$\int_A D = N \int_0^{2\pi} \int_0^1 \frac{1}{r} \cdot r dr d\theta = 2\pi N , \qquad (2.30)$$

and the integrated density:

$$\int_A \frac{1}{D} = N \int_0^{2\pi} \int_0^1 r \cdot r dr d\theta = 2\pi N /3. \qquad (2.31)$$

The efficiency of a radial sequence compared to a 2D Cartesian sequence is therefore:

$$\eta_{2D,Radial} = \frac{(\pi N)}{\sqrt{(2\pi N) \cdot \left(\frac{2\pi N}{3} \right)}} = \frac{\sqrt{3}}{2} . \qquad (2.32)$$

2.3.4   2D SPIRALS

Spiral encoding can be considered as a type of EPI encoding, as it is possible to sample the entire k-space from a single excitation as is done in single-shot EPI. The disadvantage of single-shot methods is that they are prone to both susceptibility and other off-resonance artifacts that become prominent with long readouts, such as the signal from fat. Therefore spirals, like EPI, are often acquired in multiple shots or interleaves. The number of shots dictating by how much each interleaf trajectory is rotated. The basic spiral trajectory is based upon the mathematics of an Archimedean spiral.

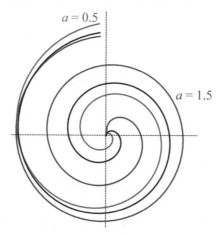

Figure 2.8: A 2D Archimedean spiral can have uniform sampling when $a = 1$ (black, middle spiral) in Equations (2.31) and (2.32). Non-uniform sampling can be created by varying the value of a. This is useful for reaching the edges of k-space more quickly ($a>1$), for when tissue susceptibility mismatch occurs, or for sampling the central portions of k-space more densely ($a<1$).

Archimedes, a Greek mathematician and physicist who lived around 250 B.C., introduced many concepts regarding areas and volumes, deriving approximations of π, and anticipating calculus with the introduction of infinitesimals and the area under a parabola. The equations for an Archimedean spiral k-space trajectory are

$$k_x = k_{max} \theta^a \sin\theta , \tag{2.33}$$

$$k_y = k_{max} \theta^a \cos\theta . \tag{2.34}$$

f $a \neq 1$, then the spiral is considered to have a variable density. With $a>1$, the spiral samples the center of k-space more densely and reaches the outer limits much more slowly (Figure 2.8). Conversely, with $a<1$, the spiral has a reduced central sampling density and higher outer density. A reduced central density is useful where T_2^* decay causes image blurring from susceptibility differences.

For a uniform density single-shot 2D spiral:

$$k_x = \frac{N}{FOV} \theta \sin\theta , \tag{2.35}$$

$$k_y = \frac{N}{FOV} \theta \cos\theta . \tag{2.36}$$

For the general solution differentiating the Archimedean spiral and converting to gradient units (where $k_{max} = N/FOV = \Delta x$), we obtain the following gradient fields:

$$G_x = \frac{2\pi}{\gamma} k_{max} \dot{\theta} \left[a\theta^{a-1} \sin\theta + \theta^a \cos\theta \right], \tag{2.37}$$

$$G_y = \frac{2\pi}{\gamma} k_{max} \dot{\theta} \left[a\theta^{a-1} \cos\theta - \theta^a \sin\theta \right]. \tag{2.38}$$

These gradient values are limited by the maximum gradient strength of the MR system. Differentiating a second time and combining x and y into a complex coordinate system to simplify derivation, we obtain the slew rate:

$$S = \frac{2\pi}{\gamma} k_{max} e^{i\theta} \left[(\ddot{\theta} - \theta\dot{\theta}^2) + i(2\dot{\theta}^2 + \theta\ddot{\theta}) \right]. \tag{2.39}$$

This slew rate must remain below the maximum slew rate limit of the gradient sub-system.

Further variations of the spiral include: TWIRL [2-11], which incorporates a radial readout followed by an Archimedean spiral; WHIRL [2-12], which uses a non-Archimedean spiral; and a spiral in, spiral out encoding for two acquisitions with a delayed TE [2-13].

2.3.5 GOLDEN ANGLE ENCODING

Rotating a k-space spoke or interleaf by the golden angle ensures a uniform, but non-repeating distribution of sampling points. Continuously acquiring data with a golden angle rotation increment will allow the reconstruction of either multiple, temporally resolved, low-resolution images or multiple interleafs can be combined into a single higher-resolution image. Typically, the golden angle rotation is applied to a radial sequence, although it is easily applied to any other 2D rotation using the rotation matrices defined later in this chapter (Figure 2.9).

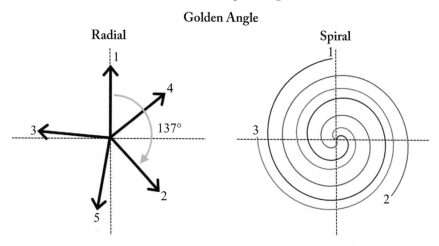

Figure 2.9: Examples of the golden angle for rotating a radial spokes (left) and interleaved spiral spokes (right). In each case, the sampled pattern would sample *k*-space along *spoke* 1, followed by sampling along *spoke* 2, then *spoke* 3, etc. Golden angle rotations is one method to reduce coherence between adjacent spokes/interleaves.

The golden angle produces incoherent signals and can be considered as a random trajectory, which is important for compressed sensing (see Chapter 4). This pseudorandom rotation has an advantage over random rotations in that less book-keeping is required. A fully random pattern requires the user to track the entire random pattern used, instead of the integer multiplied by the golden angle.

From the perspective of *k*-space sampling efficiency, a golden angle radial trajectory will always be less efficient than evenly distributed angles for a given number of spokes. The advantage of the golden angles is that its efficiency is more evenly distributed for an arbitrary number of *k*-space points. For practical imaging of real samples, golden angle rotations also reduce the phase bias from any periodic phase contributions, e.g., motion.

The golden ratio (G.R.) is the ratio of two sequential elements of the Fibonacci sequence [2-18], i.e., [1, 1, 3, 5, 8, 13, 21,...] at infinite limit. The Fibonacci sequence, *Fb*, can be generally represented as

$$F_b(n+1) = F_b(n) + F_b(n-1),$$

(2.40)

with initial conditions that $F_b(1) = F_b(2) = 1$. The Fibonacci series converges with increasing n, even though n increases to infinity. $G.R.$ is defined here as the large number ratio of two sequential elements: $G.R. = F_b(n + 1)/F_b(n)$. By dividing $F_b(n)$, this becomes

$$G.R. = \frac{F_b(n+1)}{F_b(n)} = 1 + \frac{F_b(n-1)}{F_b(n)} = \frac{F_b(n-1) + F_b(n)}{F_b(n)}. \tag{2.41}$$

At the large number limit, $F_b(n-1)/F_b(n) = F_b(n)/F_b(n+1)$:

$$G.R. = \frac{F_b(n+1)}{F_b(n)} = \frac{F_b(n-1) + F_b(n)}{F_b(n)} = \frac{F_b(n) + F_b(n+1)}{F_b(n+1)}. \tag{2.42}$$

The solution to this is

$$\text{Golden Ratio} = \frac{1 + \sqrt{5}}{2} = 1.618034 \ \text{(to 6 d.p.)} \tag{2.43}$$

This number has two important features related to its modulus (which is the fractional element after the decimal point, i.e., 0.618034): the modulus is periodic and covers all possible values between 0 and 1 when n approaches infinity. These features enable any number of n to cover all range of values within k-space, with increasing coverage for increasing n.

The Golden Angle is given by

$$\phi_{Golden} = 2\pi \ \text{mod} \left(\frac{1 + \sqrt{5}}{2} \right) = 2\pi \cdot 0.618034, \tag{2.44}$$

which simplifies to

$$\phi_{Golden} = 137.51°. \tag{2.45}$$

2.4   NON-CARTESIAN 3D ENCODING

3D encoding has several advantages over its 2D counterparts: first, if using non-selective excitations, the pulse length required to excite a 3D slab is shorter than that for a 2D slice which enables shorter TEs; second, a 3D acquisition results in a rectangular slice profile, whereas the slice profile in a 2D acquisition is non-rectangular; third, 3D encoding can obtain thinner slices than 2D, by removing bandwidth limitations in slice excitation; finally, 3D encoding enables higher SNR than 2D encoding, because all k-space readouts contribute to the SNR.

However, the advantages of 3D encoding come with limitations: first, it requires more time to acquire than 2D encoding to eliminate image aliasing in the second slice-encoding direction; secondly, ththis second-phase encoding, while bringing an extra degree of freedom, increases the difficulty of optimization problems; and thirdly, the second-phase encoding requires additional computational power for advanced reconstructions. Simple tricks can enable fast processing, such as reducing the reconstruction matrix and combining any temporal points using a temporal subspace reconstruction (see Chapter 4).

2.4.1   3D RADIAL ENCODING

Expanding from 2D radial encoding, a 3D radial sequence has the pattern:

$$G_x = G_0 \sin\theta\cos\phi,$$ (2.46)

$$G_y = G_0 \cos\theta\cos\phi,$$ (2.47)

$$G_z = G_0 \sin\phi,$$ (2.48)

where ϕ and θ vary at the beginning of each spoke.

With uniform sampling, the efficiency of a 3D radial sequence can be calculated similarly, using a density of $1/r^2$ and spherical volume integral:

$$\eta_{3D,Radial} = \frac{\left(4\pi N/3\right)}{\sqrt{\left(4\pi N\right)\cdot\left(\frac{4\pi N}{5}\right)}} = \frac{\sqrt{5}}{3}.$$ (2.49)

Therefore, we can multiply the lines required by our standard 3D Cartesian matrix by $\frac{\sqrt{5}}{3}$ to obtain the number of spokes that fulfill the Nyquist limit.

In order to fully sample a 3D sphere, several orderings of trajectory interleaves, or (θ,ϕ) combinations can be used. For instance, uniform sampling [2-19] can be employed, moving slowly from the top to the bottom of the sphere. For N spokes, the ith spoke will have spherical angles θ and ϕ for the calculation of G_x and G_y in Equations 2.46 and 2.47. The values of G_z, $\theta(i)$ and $\phi(i)$ is then calculated from:

$$G_z(i) = -\frac{i-1}{N-1}, \qquad 1 \leq i \leq N,$$ (2.50)

$$\phi(i) = \arccos\left(G_z(i)\right),$$ (2.51)

$$\theta(i) = \left(\theta(i-1) + \frac{3.6}{\sqrt{N\left(1-G_z^2\right)}}\right)\mod\left(2\pi\right), \qquad 2 \leq i \leq N-1.$$ (2.52)

This trajectory is predictable, as the user only has control of N, which enables simple book-keeping for reconstruction (which is an important point for non-Cartesian encoding). While uniform, the pattern is highly ordered and increases image artifacts from any regular temporal pattern, which re-duces the ability to use an arbitrary number of undersampled points for reconstruction.

Chan [2-20] proposed two ratios for use in 3D imaging with improved undersampling tolerance—similar to the 2D golden angle. These are called the golden means and have the values:

$$f_1 = 0.6823, \tag{2.53}$$

$$f_2 = 0.4656. \tag{2.54}$$

In order to distribute the density across the sphere in the slice direction, the second rotational angle, ϕ_{gold}, has its density related to its arccosine. Thus, for the ith readout, the angles are:

$$\theta(i) = i \cdot f_1 \cdot 2\pi, \tag{2.55}$$

$$\phi(i) = \arccos(i \cdot f_2). \tag{2.56}$$

2.4.2   ADVANCED 3D ENCODING

3D radial sampling is not efficient for covering k-space compared to any method that acquires data in more than a single line per excitation. The simplest 3D encoding method is to stack a 2D acquisition, two other common methods are 3D twisted projection imaging (TPI) and cones (3D spirals) [2-21–2-23] (see Figure 2.10).

3D stacking includes the stack of stars, and stack of spirals [2-22]. To achieve 2D stacking, a 2D non-Cartesian k-space trajectory is repeated using a conventional phase-encoding gradient in the slab select direction. With a slightly more advanced form, the 2D trajectory is repeated with a 2D rotation in order to reduce the temporal coherence between adjacent slices. Temporal coherence can also be reduced through non-sequential ordering of the slice-phase.

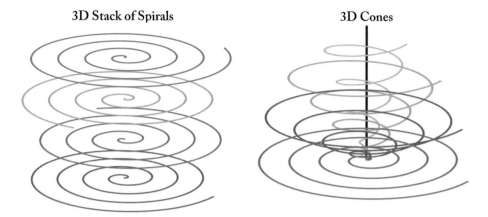

3D Stack of Spirals   **3D Cones**

Figure 2.10: Two spiral-based 3D methods are shown: a stack of spirals (left) and 3D spirals / cones (right). The stack of spirals is shown with each spiral rotated by the golden angle to reduce sampling coherence between slices. The 3D cones sequence consists of spirals that have a changing amplitude and width to sample a spherical volume. The 3D cones sequence is shown with only the positive half of k-space for clarity, although this would be repeated for a symmetric negative half to increase the stability from off-resonance effects.

2.4.3   3D EULER ROTATIONS

2D trajectories can fill 3D k-space with rotated, instead of stacked, repeats of the trajectory. The complication with this method is that unlike in two dimensions, rotations in three dimensions do not commute, hence the order of rotations matters. 3D rotations use standard rotation matrices, similar to those discussed in Chapter 3.

A standard convention of Euler rotations is "yaw," "pitch," and "roll," which refers to the rotations about the z, y, and x axes, and where $R_z(\psi)$, $R_y(\theta)$, and $R_x(\phi)$ refer to their respective rotation matrices. These are also known as the normal/precession, transversal/nutation, and longitudinal/intrinsic rotations, respectively. These are often, but not always, performed in the order of "roll" (x) first, "pitch" (y) second, and "yaw" (z) third, i.e., $R_z(\psi)R_y(\theta)R_x(\phi)$, particularly when acting on row vectors. These are also known as the Tait-Bryan, nautical, or Cardan rotations.

The full rotation matrix in order of $R_z(\psi)R_y(\theta)R_x(\phi)$ is considered the intrinsic rotation, as follows:

$R_z(\psi)R_y(\theta)R_x(\phi) =$

$$
\begin{bmatrix}
\cos\theta\cos\psi & -\cos\phi\sin\psi+\sin\phi\,\sin\theta\cos\psi & \sin\phi\sin\psi+\cos\phi\sin\theta\cos\psi \\
\cos\theta\sin\psi & \cos\phi\cos\psi+\sin\phi\sin\theta\,\sin\psi & -\sin\phi\cos\psi+\cos\phi\sin\theta\,\sin\psi \\
-\sin\theta & \cos\theta\sin\phi & \cos\theta\cos\phi
\end{bmatrix}. \quad (2.57)
$$

When performing a pre-multiplication of the rotation matrix on a columnar vector, we use the extrinsic rotation, which occurs in the opposite order, $R_x(\phi)R_y(\theta)R_z(\psi)$, such that:

$$
\begin{bmatrix} k_x \\ k_y \\ k_z \end{bmatrix} =
$$

$$
\begin{bmatrix} \cos\theta\cos\psi & -\cos\theta\sin\psi & \sin\theta \\ -\cos\phi\sin\psi+\sin\phi\cos\psi & \cos\phi\cos\psi-\sin\phi\sin\theta\sin\psi & -\cos\theta\sin\phi \\ \sin\phi\sin\psi-\cos\phi\sin\theta\cos\psi & \sin\phi\cos\psi+\cos\phi\sin\theta\sin\psi & \cos\theta\cos\phi \end{bmatrix} \cdot \begin{bmatrix} k_{x,0} \\ k_{y,0} \\ k_{z,0} \end{bmatrix} \cdot \tag{2.58}
$$

For two angles that consider only roll, ϕ, and pitch, θ, such that $\psi = 0$, the extrinsic rotation matrix, $R_x(\phi)R_y(\theta)$, rotates a k-space vector with:

$$
\begin{bmatrix} k_x \\ k_y \\ k_z \end{bmatrix} = \begin{bmatrix} \cos\theta\cos\psi & -\cos\theta\sin\psi & \sin\theta \\ -\sin\psi & \cos\psi & 0 \\ \sin\theta\cos\psi & \sin\theta\sin\psi & \cos\theta \end{bmatrix} \cdot \begin{bmatrix} k_{x,0} \\ k_{y,0} \\ k_{z,0} \end{bmatrix} . \tag{2.59}
$$

2.5 PHASE OFFSETS

2.5.1 PARTIAL FOURIER ENCODING

With ideal sampling, k-space has symmetry about any axis, and can be mirrored with the phase-corrected complex conjugate of the mirrored axis (called homodyne detection [2-24]). This allows us to reduce the amount of data required for encoding by obtaining just over half the data for similar reconstruction. However, partial Fourier encoding is sensitive to phase offsets from motion and off-resonance. Cartesian partial encoding will often undersample by 5/8—where 4/8 is only half of k-space sampling—to ensure phase stability and easier reconstruction.

2.5.2 MOTION

Bulk motion, such as from the heart, respiration, or patient movement, changes the phase of individual k-space acquisitions and results in image artifacts. Both contrast encoding and motion can introduce phase changes in the k-space center. Fast encoding schemes, such as EPI, and k-space centric encoding schemes, such as radial, are less sensitive to motion than spin warp encoding.

With non-Cartesian schemes, the effects of motion are reduced due to the directionality of encoding and due to the high amount of central k-space oversampling. Non-Cartesian encoding measures k-space across many different spatial directions—some opposite—reducing the impact of motion from any singular direction. In-plane motion can result in artifact, like the undersampled *FOV* artifact, presenting as increased image speckling. Through-plane motion can result in additional artifacts, particularly if the tissue moves through the imaging plane, e.g., respiration, during data acquisition.

2.5.3   OFF-RESONANCE

Differences in resonant frequency, i.e., off-resonance from water due to air or fat can cause image blurring for long encoding times. Fast encoding reduces image blurring from off-resonance and relaxation effects, which often results in reduced *SNR* due to the higher sampling bandwidth required.

Considering water alone, the signal will be modulated during *k*-space readout by the off-resonance ΔB_0:

$$S_{H20}\left(k_x(t), k_y(t)\right) = S_{0,H20}\left(k_x(t), k_y(t)\right)\exp\left(i\gamma t\, \Delta B_0\right). \tag{2.60}$$

Fat also has an off-resonance Δf_{fat} with a similar, but separate, signal modulation:

$$S_{fat}\left(k_x(t), k_y(t)\right) = S_{0,fat}\left(k_x(t), k_y(t)\right)\exp\left(i\gamma t\left[\Delta B_0 + \Delta f_{fat}\right]\right). \tag{2.61}$$

These will add linearly when both are present within a voxel, such as at tissue boundaries. Replacing *t* with *TE* and readout time, $T_{readout}$, the signal will be modulated to

$$S\left(k_x(t), k_y(t)\right) = S_0\left(k_x(t), k_y(t)\right)\exp\left(i\gamma\left(TE + T_{readout}\right)\cdot\left[\Delta B_0 + \Delta f\right]\right). \tag{2.62}$$

This formulation allows us to estimate the effect of the off-resonance artifact in terms of a spatial shift of the signal,

$$Off-resonance\ Shift \propto \left[\Delta B_0 + \Delta f\right]\cdot\left(TE + T_{readout}\right). \tag{2.63}$$

In addition, the electron structure of water and fat means that there is a small shift in their resonant frequencies (Δf_{fat} = 3.5 ppm), referred to as the chemical shift. The number of voxels shifted in image space due to this chemical shift is

$$\Delta x_{voxels} = \frac{1}{\Delta f_{fat}} \cdot \frac{N_{matrix}}{BW_{receiver}}. \tag{2.64}$$

Thus, a higher receiver bandwidth reduces off-resonance artifact and fat/water chemical shift artifact.

2.6   CONCLUSION

In this chapter, we discussed how to encode several types of trajectories, and some of the underlying principles for deciding on those trajectories. The trajectory decision is influenced by many factors, such as scan time, field-of-view, resolution, motion, and available *SNR*. With fast encoding methods, particularly those that enable high incoherence following multiple readouts, such as the non-Cartesian methods discussed, we have the potential to encode temporal signals, such as time-varying image contrast, which we will discuss in the next chapter, more efficiently. The next chapter will discuss how contrast may be encoded in the MR signals.

BIBLIOGRAPHY

[2-1]   P. C. Lauterbur, Image formation by induced local interactions: examples employing nuclear magnetic resonance, *Nature*, 246, p. 469, 1974. 17, 24, 25, 26

[2-2]   P. Mansfield and P. K. Grannell, NMR 'diffraction' in solids?, *J. Phys. C Solid State Phys.*, 6(22) pp. L422–L426, Nov. 1973. DOI: 10.1088/0022-3719/6/22/007. 17

[2-3]   P. C. Lauterbur, Magnetic resonance zeugmatography, *Pure Appl. Chem.*, 40(1–2), pp. 149–157, Jan. 1974. DOI: 10.1351/pac197440010149. 17

[2-4]   P. C. Lauterbur, D. M. Kramer, W. V. House, and C.-N. Chen, Zeugmatographic high resolution nuclear magnetic resonance spectroscopy. Images of chemical inhomogeneity within macroscopic objects, *J. Am. Chem. Soc.*, 97(23), pp. 6866–6868, Nov. 1975. DOI: 10.1021/ja00856a046. 17

[2-5]   P. Mansfield, Multiplanar image formation using NMR spin echoes, *J. Phys. C Solid State Phys.*, 10, pp. 850–855, 1981. 17

[2-6]   A. Kumar, D. Welti, and R. R. Ernst, NMR Fourier zeugmatography, *J. Magn. Reson.*, 18(1), pp. 69–83, Apr. 1975. DOI: 10.1016/0022-2364(75)90224-3. 17

[2-7]   W. A. Edelstein, J. M. S. Hutchison, G. Johnson, and T. Redpath, Spin warp NMR imaging and applications to human whole-body imaging, *Phys. Med. Biol.*, 25(4), pp. 751–756, Jul. 1980. DOI: 10.1088/0031-9155/25/4/017. 17

[2-8]   G. H. Glover and J. M. Pauly, Projection reconstruction techniques for reduction of motion effects in MRI, *Magn. Reson. Med.*, 28(2), pp. 275–289, Dec. 1992. DOI: 10.1002/mrm.1910280209. 24, 25, 26

[2-9]   D. C. Peters et al., Undersampled projection reconstruction for active catheter imaging with adaptable temporal resolution and catheter-only views, *Magn. Reson. Med.*, 49(2), pp. 216–222, Feb. 2003. DOI: 10.1002/mrm.10390. 24, 25, 26

[2-10]   D. C. Peters et al., Undersampled projection reconstruction applied to MR angiography," *Magn. Reson. Med.*, 43(1), pp. 91–101, Jan. 2000. DOI: 10.1002/(SICI)1522-2594(200001)43:1<91::AID-MRM11>3.0.CO;2-4. 24, 25, 26

[2-11]   J. I. Jackson, D. G. Nishimura, and A. Macovski, Twisting radial lines with application to robust magnetic resonance imaging of irregular flow, *Magn. Reson. Med.*, 25(1), pp. 128–139, May 1992. DOI: 10.1002/mrm.1910250113. 24, 30

[2-12]   J. G. Pipe, An optimized center-out *k*-space trajectory for multishot MRI: Comparison with spiral and projection reconstruction, *Magn. Reson. Med.*, 42(4), pp. 714–720, Oct. 1999. DOI: 10.1002/(SICI)1522-2594(199910)42:4<714::AID-MRM13>3.0.CO;2-G. 24, 30

[2-13]   G. H. Glover and C. S. Law, Spiral-in/out BOLD fMRI for increased *SNR* and reduced susceptibility artifacts, *Magn. Reson. Med.*, 46(3), pp. 515–522, Sep. 2001. DOI: 10.1002/mrm.1222. 24, 30

[2-14]   H. H. Wu, J. H. Lee, and D. G. Nishimura, MRI using a concentric rings trajectory, *Magn. Reson. Med.*, 59(1), pp. 102–112, Jan. 2008. DOI: 10.1002/mrm.21300. 24

[2-15]   W. S. Hinshaw, Image formation by nuclear magnetic resonance: The sensitive-point method, *J. Appl. Phys.*, 47(8), pp. 3709–3721, Aug. 1976. DOI: 10.1063/1.323136. 24

[2-16]   G. E. Sarty, Single TrAjectory Radial (STAR) imaging, *Magn. Reson. Med.*, 51(3), pp. 445–451, Mar. 2004. DOI: 10.1002/mrm.20001. 24

[2-17]   D. C. Noll, S. J. Peltier, and F. E. Boada, Simultaneous multislice acquisition using rosette trajectories (SMART): A new imaging method for functional MRI, *Magn. Reson. Med.*, 39(5), pp. 709–716, May 1998. DOI: 10.1002/mrm.1910390507. 24,

[2-18]   A. F. Horadam, A generalized Fibonacci sequence, *Am. Math. Mon.*, 68(5), pp. 455–459, May 1961. DOI: 10.1080/00029890.1961.11989696. 31

[2-19]   E. B. Saff and A. B. J. Kuijlaars, Distributing many points on a sphere, *Math. Intell.*, 19(1), pp. 5–11, Dec. 1997. DOI: 10.1007/BF03024331. 33

[2-20]   R. W. Chan, E. A. Ramsay, C. H. Cunningham, and D. B. Plewes, Temporal stability of adaptive 3D radial MRI using multidimensional golden means, *Magn. Reson. Med.*, 61(2), pp. 354–363, Feb. 2009. DOI: 10.1002/mrm.21837. 34

[2-21]   F. E. Boada, G. X. Shen, S. Y. Chang, and K. R. Thulborn, Spectrally weighted twisted projection imaging: ReducingT2 signal attenuation effects in fast three-dimensional sodium imaging, *Magn. Reson. Med.*, 38(6), pp. 1022–1028, Dec. 1997. DOI: 10.1002/mrm.1910380624. 34

[2-22]   P. Irarrazabal and D. G. Nishimura, Fast three dimensional magnetic resonance imaging, *Magn. Reson. Med.*, 33(5), pp. 656–662, May 1995. DOI: 10.1002/mrm.1910330510. 34

[2-23]   P. T. Gurney, B. A. Hargreaves, and D. G. Nishimura, Design and analysis of a practical 3D cones trajectory, *Magn. Reson. Med.*, 55(3), pp. 575–582, Mar. 2006. DOI: 10.1002/ mrm.20796. 34

[2-24]   D. C. Noll, D. G. Nishimura, and A. Macovski, Homodyne detection in magnetic resonance imaging, *IEEE Trans. Med. Imaging*, 10(2), pp. 154–163, 1991. DOI: 10.1109/42.79473. 36

CHAPTER   3

Contrast Encoding

In Chapters 1 and 2, we discussed the basics of NMR spin dynamics and spatial encoding respectively. Here we discuss basic pulse sequences and derive the equations for the partitioning of magnetization into configurations. Then, we present the concept of the extended phase graph method to illustrate the effects of RF pulses and relaxation. Finally, we discuss qMRI methods to acquire data with different T_1 and/or T_2 contrasts in order to fit a quantitative model.

3.1   PULSE SEQUENCES

In MRI, pulse sequences are temporal combinations of RF and gradient pulses. The RF pulses rotate the longitudinal and/or transverse magnetization while the gradient pulses have the effect of dephasing the magnetization and are generally used also to spatially localize the signal in order to create an image. Spatial localization was discussed in Chapter 2, while here we focus on the flip angles and timings between RF and how gradient pulses are manipulated in order to generate the contrast in an MRI image and, if desired, make an estimate of the T_1 and T_2 relaxation times. The following section provides a brief introduction to the main strategies used to manipulate nuclear spins in order to get meaningful MRI signals producing contrasts between different biological tissues.

3.1.1   FREE INDUCTION DECAY

The simplest pulse sequence is a single RF excitation pulse that tips the longitudinal magnetization by a given flip angle, resulting in the creation of detectable transverse magnetization known as a free induction decay (FID). As discussed, in Section 1.6 the transverse magnetization will dephase with a time constant called T_2^*. The time between subsequent RF excitation pulses, during which the magnetization will recover due to T_1 relaxation, is referred to as the repetition time (TR). The direct acquisition of FID signals is often used in magnetic resonance spectroscopy (MRS), however in MRI the FID is usually collected as part of a gradient echo acquisition described below.

3.1.2   SPIN ECHO

Since the FID decays due to T_2^* relaxation a spin echo sequence is used to eliminate the T_2' contribution resulting in a pure T_2 signal contribution. The original spin echo sequence comprised of two 90° pulses, separated by a time τ [3-1]. The first 90° pulse created transverse magnetization that dephases due to T_2^*. The second 90° pulse rephases a component of this magnetization resulting in the classic Hahn or so-called "eight-ball" echo at a time 2τ after the first pulse, i.e., [90° – τ – 90° – τ – Hahn_echo – T_1 recovery] (Figure 3.1). The effect of this refocusing is to reverse the fixed dephasing due to T_2'. However, the dephasing caused by intrinsic T_2 effects is irreversible, hence the signal will only be attenuated by this mechanism. The time between the first (excitation) pulse and the peak of the echo signal is known as the echo time (TE), i.e., $TE = 2\tau$. The time between subsequent excitation pulses is the TR.

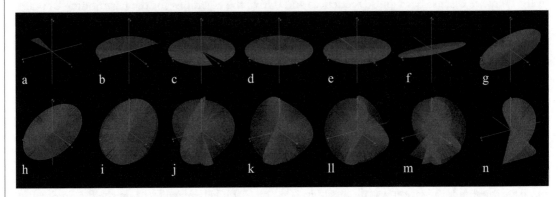

Figure 3.1: Vector representation of the classic Hahn 8-ball echo. Following the first 90° RF pulse the magnetization starts to dephase in the transverse (x–y) plane (a–e), The second 90° RF pulse about the x-axis then rotates this disk of dephased magnetization into the x–z plane (f–i). The magnetization then continues to precess (j–m), eventually forming the 8-ball echo (n). Figure created using the Interactive Spin Viewer developed by Dr. Stefan Petersson, GE Healthcare, Sweden.

In practice, it is more common to use a spin echo sequence comprising a 90°, 180° pulse pair, i.e., [90° – τ – 180° – τ – spin_echo – T_1 recovery]. The initial transverse magnetization is fully dephased in the transverse plane before the application of the 180° pulse that flips the entire magnetization resulting in a natural rephasing in the transverse plane at a time $TE = 2\tau$ after the initial 90° pulse (Figure 3.2). This results in an echo signal twice the magnitude of the Hahn echo for the same TE. The signal equation for such a spin echo sequence is:

$$S = S_0 \left(1 - e^{-\frac{TR}{T_1}} \right) e^{-\frac{TE}{T_2}}, \tag{3.1}$$

where S_0 is the equilibrium signal. The intrinsic difference in tissue T_1, T_2 and proton density will result in different signal amplitudes, and hence contrast between tissues, depending upon the chosen TE and TR. The transverse magnetization will decay during the TE period, while the longitudinal magnetization will recover during TR. Due to practical limitations on TE and TR the images from standard spin echo sequences are referred to as being contrast "weighted" since the signal is not purely from a single tissue parameter.

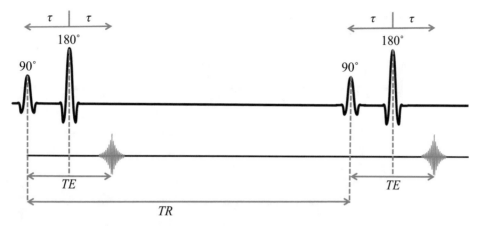

Figure 3.2: Basic 90°–180° spin echo sequence. The echo forms at a time *TE* after the 90° excitation pulse, which is equal to twice the time, τ, between the 90° and 180° pulses. The sequence is repeated with a repetition time (*TR*), allowing for T_1 recovery.

3.1.3   STIMULATED ECHO

Stimulated echoes are not widely used for MR imaging but are often used for MR spectroscopy. However, they can contribute to the signal formation in several pulse sequences. A stimulated echo arises from three RF pulses (Figure 3.3). The first pulse creates a component of transverse magnetization, the second RF pulse tips a component of that transverse magnetization into the longitudinal direction for a period of time, sometimes called the mixing time (*TM*), where it does not dephase any further but recovers due to T_1 relaxation, until the third RF pulse tips a component of this longitudinal magnetization back into the transverse plane. The net effect is to cause refocusing of the transverse magnetization and the formation of a stimulated echo, i.e., for three α pulses [$\alpha - \tau - \alpha - TM - \alpha - \tau -$ stimulated_echo $- T_1$ recovery]). If $TM = \tau$ then the stimulated echo will occur at the same time as the Hahn echoes from the preceding RF pulses introducing an additional T_1-weighting into the signal.

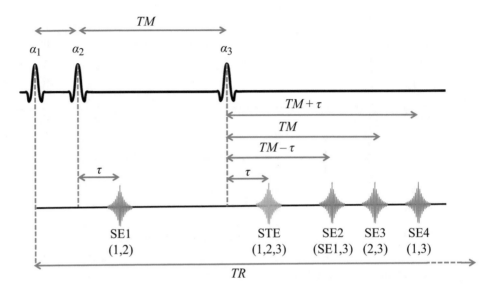

Figure 3.3: Evolution of a stimulated echo (STE) from three α pulses, separated by time τ and TM. Note that in addition to the STE four Hahn echoes can also occur. Three echoes arise from the refocusing of two RF pulses (SE1, SE3, and SE4), while a fourth echo (SE2) arises from a second refocusing of the SE1 echo by the third α pulses. The sequence is repeated with a repetition time (TR).

3.1.4   INVERSION RECOVERY

If we apply a $180°$ pulse to the initial magnetization M, aligned along the z-direction, it will become inverted. The magnetization will then recover exclusively due to $T1$ relaxation, since no transverse magnetization has been created. Following a suitable time period, known as the inversion time (TI), the recovered z-magnetization is tipped into the transverse plane with the signal typically being formed by a spin echo, i.e., $[180° - TI - 90° - \tau - 180° - \tau - \text{spin\_echo} - T1 \text{ recovery}]$ (Figure 3.4). The inversion recovery sequence can be used to increase the $T1$-weighting within an image or more commonly to null specific tissues, e.g., fat. Acquiring images with different TI values can be used to quantify $T1$ relaxation.

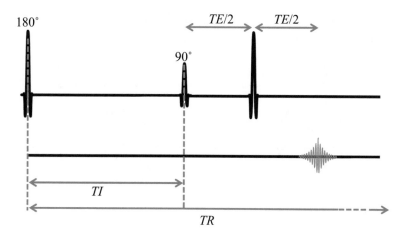

Figure 3.4: Basic inversion recovery spin echo sequence. The initial magnetization is inverted by the first 180° pulse. After the chosen inversion time (*TI*) the signal is formed by a spin echo. The sequence is repeated with a repetition time (*TR*) to allow T_1 recovery.

3.1.5 GRADIENT ECHO

A gradient echo sequence forms an echo signal through the application of a magnetic field gradient pulse that forcibly dephases the transverse magnetization and then forcibly rephases it by a gradient of equal area but opposite polarity. For an introduction to gradient pulses and their use in spatial encoding, see Chapter 2.

Since an RF refocusing pulse is not used, the *TE* can be shorter than a spin echo sequence, however the echo amplitude is dependent on T_2^* rather than T_2 decay. Gradient echo sequences also typically use RF excitation pulses $\alpha < \frac{\pi}{2}$ (90°), which allows the use of a shorter *TR* while still maintaining an acceptable signal amplitude, i.e., $[\alpha - \tau - \text{gradient\_echo} - T_1 \text{ recovery}]$. A consequence of a shorter *TR* is that there may still be appreciable transverse magnetization at the time of the next RF pulse, e.g., if $TR << T_2$. This second pulse may then act as a refocusing pulse, i.e., the *TR* effectively becomes τ in the spin echo description above, adding a T_2-weighted component into the detected signal. There are three main types of gradient echo sequence. The first is the ("fully balanced") steady-state free precession sequence (SSFP) in which the imaging gradient areas are balanced on all three axes (Figure 3.5a). A limitation of this sequence is that it is very sensitive to magnetic field non-uniformities. A slightly different variant of this sequence ("gradient spoiled") doesn't balance all the gradients resulting in a constant dephasing of the residual transverse magnetization and a reduced sensitivity to field non-uniformities (Figure 3.5b). There is also a third type of gradient echo sequence, "RF spoiled," that destroys the residual transverse magnetization using an appropriate RF phase cycling scheme. The various MRI system vendors call these gradient echo variants by different names, Table 3.1 provides a basic list of vendor acronyms.

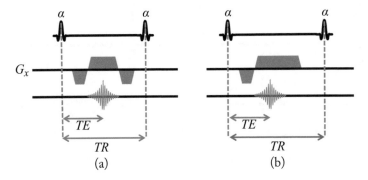

Figure 3.5: Gradient echo sequences: (a) the bSSFP sequence has all the gradients fully balanced on each axis, so the time integral, or first moment, of the gradient fields for each TR is zero. This diagram only shows the G_x frequency encoding gradient, but both the slice select and phase encoding gradients are also fully balanced; (b) the gradient-spoiled and RF-spoiled sequence have unbalanced gradients, so the integral of the gradient fields is different from zero, but equal for each TR. The only difference between the gradient-spoiled and RF-spoiled version is that in the RF-spoiled version the excitation pulses α are phase cycled to destroy the residual transverse magnetization.

Table 3.1. Gradient echo sequence acronyms. FE: Field Echo, FFE: Fast Field Echo, FIESTA: Fast Imaging with Enhanced Steady sTate Acquisition, FISP: Fast Imaging with Steady Precession, FLASH: Fast Low Angle SHot, GRE: Gradient Recalled Echo, GRASS: Gradient Recalled Acquisition in the Steady State, SPGR: SPoiled GRass, SARGE: Steady state Acquisition Rewound Gradient Echo

Generic	Philips	Siemens	GE	Hitachi	Toshiba
Gradient spoiled	FFE	FISP	GRE (was GRASS)	Rephased SARGE	SSFP
RF spoiled	T1-FFE	FLASH	SPGR	RF Spoiled SARGE	T1-FFE
Balanced steady-state free precession	Balanced-FFE	TrueFISP	FIESTA	Balanced SARGE	True SSFP

3.1.6   FAST/TURBO SPIN ECHO

Fast/turbo spin echo (FSE/TSE) sequences use an excitation pulse followed by multiple RF refocussing pulses to create a train of spin echoes (Figure 3.6), i.e., [$90° - \tau - 180° - \tau -$ spin_echo $- \tau - 180° - \tau -$ spin_echo $- ... - T_1$ recovery]. While multiple spin echoes can be used to quantify T_2 relaxation the original Fast Spin Echo (FSE)/Turbo Spin Echo (TSE) method, known as Rapid

Acquisition with Relaxation Enhancement (RARE) [3-2], was used to accelerate image acquisition. Each spin echo was individually spatially encoded and incorporated into the raw data for the whole image (see Chapter 2). More recently, there has been considerable interest in reducing and/or varying the amplitude of the refocusing pulses across the train of echoes, to improve image quality and to manage the RF power deposition associated with large RF flip angles, especially at higher magnetic field strengths.

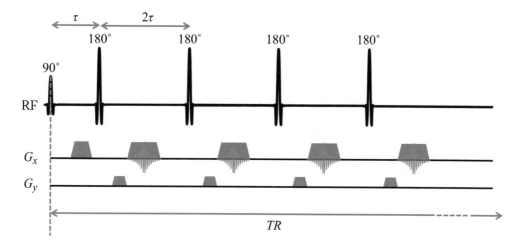

Figure 3.6: Scheme of a fast spin echo, or RARE sequence with an ETL of 4. Refocusing pulses are spaces 2τ apart and an echo forms at every 2τ after the 90° excitation pulse. The sequence is repeated with a repetition time (TR) to allow T_1 recovery.

3.2   BLOCH EQUATION SIMULATIONS

The most accurate method to investigate the effect of a pulse sequence on the magnetization is to perform a discrete time simulation of the Bloch equations. This involves the use of rotation matrices to describe the effect of the RF pulses about a given axis of rotation and exponential terms to describe the relaxation of the magnetization. The simulations also assume signal dephasing and rephasing. While this may arise due to intrinsic magnetic field non-uniformities it is usual to associate these effects with the imaging gradients. We will start with a brief review of rotations using matrices.

A rotation of angle θ about the x-axis can be expressed as

$$R_x(\theta) = \begin{pmatrix} 1 & 0 & 0 \\ 0 & \cos(\theta) & -\sin(\theta) \\ 0 & \sin(\theta) & \cos(\theta) \end{pmatrix}.$$

(3.2)

A rotation of angle θ about the y-axis can be expressed as

$$R_y(\theta) = \begin{pmatrix} \cos(\theta) & 0 & \sin(\theta) \\ 0 & 1 & 0 \\ -\sin(\theta) & 0 & \cos(\theta) \end{pmatrix}.$$

(3.3)

A rotation of angle θ about the z-axis can be expressed as

$$R_z(\theta) = \begin{pmatrix} \cos(\theta) & \sin(\theta) & 0 \\ -\sin(\theta) & \cos(\theta) & 0 \\ 0 & 0 & 1 \end{pmatrix}.$$

(3.4)

We can express the transverse magnetization in complex terms, i.e,

$$M_{xy} = M_x + iM_y$$

(3.5)

and its complex conjugate

$$M_{xy}^* = M_x + iM_y.$$

(3.6)

In matrix form this conversion can be performed by a basis transformation

$$M = \begin{pmatrix} M_{xy} \\ M_{xy}^* \\ M_z \end{pmatrix} = S \begin{pmatrix} M_x \\ M_y \\ M_z \end{pmatrix}, \text{ where } S = \begin{pmatrix} 1 & i & 0 \\ 1 & -i & 0 \\ 0 & 0 & 1 \end{pmatrix},$$

(3.7)

such that

$$\begin{pmatrix} 1 & i & 0 \\ 1 & -i & 0 \\ 0 & 0 & 1 \end{pmatrix} \cdot \begin{pmatrix} M_x \\ M_y \\ M_z \end{pmatrix} = \begin{pmatrix} M_x + iM_y \\ M_x - iM_y \\ M_z \end{pmatrix} = \begin{pmatrix} M_{xy} \\ M_{xy}^* \\ M_z \end{pmatrix}.$$

(3.8)

The inverse transformation of S is given by

$$S^{-1} = \frac{1}{2} \begin{bmatrix} 1 & 1 & 0 \\ 1-i & i & 0 \\ 0 & 0 & 2 \end{bmatrix},$$

(3.9)

such that

$$\frac{1}{2}\begin{bmatrix} 1 & 1 & 0 \\ 1-i & i & 0 \\ 0 & 0 & 2 \end{bmatrix} \cdot \begin{bmatrix} M_{xy} \\ M_{xy}^* \\ M_z \end{bmatrix} = \begin{pmatrix} M_x \\ M_y \\ M_z \end{pmatrix}. \tag{3.10}$$

The new transformation matrix encompassing a rotation α about the x-axis is given by $T_x(\alpha)$,

$$T_x(\alpha) = S \cdot R_x(\alpha) \cdot S^{-1} =$$

$$\begin{bmatrix} \frac{1}{2}(1+\cos(\alpha)) & \frac{1}{2}(1-\cos(\alpha)) & -i\sin(\alpha) \\ \frac{1}{2}(1-\cos(\alpha)) & \frac{1}{2}(1+\cos(\alpha)) & i\sin(\alpha) \\ \frac{-i}{2}\sin(\alpha) & \frac{i}{2}\sin(\alpha) & \cos(\alpha) \end{bmatrix}. \tag{3.11}$$

which can be rewritten using the standard trigonometrical identities $\frac{1}{2}(1 + \cos(\alpha)) = \cos^2\left(\frac{\alpha}{2}\right)$ and $\frac{1}{2}(1 - \cos(\alpha)) = \sin^2\left(\frac{\alpha}{2}\right)$ to give

$$T_x(\alpha) = \begin{bmatrix} \cos^2\left(\dfrac{\alpha}{2}\right) & \sin^2\left(\dfrac{\alpha}{2}\right) & -i\sin(\alpha) \\ \sin^2\left(\dfrac{\alpha}{2}\right) & \cos^2\left(\dfrac{\alpha}{2}\right) & i\sin(\alpha) \\ \dfrac{-i}{2}\sin(\alpha) & \dfrac{i}{2}\sin(\alpha) & \cos(\alpha) \end{bmatrix}. \tag{3.12}$$

Note that T_x does not have the property of a rotation matrix anymore because of the basis transformation.

Similarly, a rotation ϕ about the z-axis can be described by the transformation matrix $T_z(\phi)$

$$T_z(\phi) = S \cdot R_z(\phi) \cdot S^{-1}$$

$$= \begin{pmatrix} 1 & i & 0 \\ 1 & -i & 0 \\ 0 & 0 & 1 \end{pmatrix} \cdot \begin{pmatrix} \cos(\phi) & \sin(\phi) & 0 \\ -\sin(\phi) & \cos(\phi) & 0 \\ 0 & 0 & 1 \end{pmatrix} \cdot \frac{1}{2}\begin{bmatrix} 1 & 1 & 0 \\ 1-i & i & 0 \\ 0 & 0 & 2 \end{bmatrix}$$

$$= \begin{bmatrix} \cos(\phi)+i\sin(\phi) & 0 & 0 \\ 0 & \cos(\phi)-i\sin(\phi) & 0 \\ 0 & 0 & 1 \end{bmatrix} \tag{3.13}$$

$$= \begin{bmatrix} e^{i\phi} & 0 & 0 \\ 0 & e^{-i\phi} & 0 \\ 0 & 0 & 1 \end{bmatrix}.$$

Combining the above two results $T_\phi(\alpha) = T_z(\phi)\, T_x(\alpha)\, T_z(-\phi)$ gives the solution for a general RF pulse with a flip angle of α with an initial RF phase of ϕ which acts on the complex magnetization vector, such that the magnetization after the pulse (positive superscript) is related to that prior to the pulse (negative superscript) as follows:

$$\begin{bmatrix} M_{xy} \\ M_{xy}^* \\ M_z \end{bmatrix}^+ = \begin{bmatrix} \cos^2\left(\dfrac{\alpha}{2}\right) & e^{2i\phi}\sin^2\left(\dfrac{\alpha}{2}\right) & -ie^{i\phi}\sin(\alpha) \\ e^{-2i\phi}\sin^2\left(\dfrac{\alpha}{2}\right) & \cos^2\left(\dfrac{\alpha}{2}\right) & ie^{-i\phi}\sin(\alpha) \\ \dfrac{-i}{2}e^{-i\phi}\sin(\alpha) & \dfrac{i}{2}e^{i\phi}\sin(\alpha) & \cos(\alpha) \end{bmatrix} \begin{bmatrix} M_{xy} \\ M_{xy}^* \\ M_z \end{bmatrix}^-. \tag{3.14}$$

This equation demonstrates the so-called "partitioning effect" of an arbitrary RF pulse. The magnetization can be split into three parts, from the perspective of the initial transverse magnetization M_+: (i) dephasing transverse magnetization M_+; (ii) rephasing transverse magnetization M_-; and (iii) longitudinal magnetization M_z. The first component can be considered as not being affected by the RF pulse, and hence is sometimes represented as a "0°-like" pulse with a fraction proportional to $\cos^2\left(\frac{\alpha}{2}\right)$, the second component will rephase to form an echo and is therefore represented by a "180°-like" pulse with a fraction proportional to $\sin^2\left(\frac{\alpha}{2}\right)$, while the final component becomes longitudinal magnetization and is represented as a "90°-like" pulse with a fraction proportional to $\sin(\alpha)$.

The concept of partitioning magnetization into configurations was first proposed by Woessner to describe NMR diffusion experiments [3-3], and more recently reintroduced by Hennig for the description of pulse sequences which have multiple RF pulses, separated by fixed time intervals and balanced gradients [3-4].

The phase increment θ created by a gradient (G) between RF pulses separated by a time (t) is given by

$$\theta(t) = \gamma \int_0^t G(t)z\,dt, \tag{3.15}$$

where z is the location across a voxel. The phase is a simple twist in the magnetization across the voxel. The application of multiple pulses and dephasing periods can therefore be compactly represented as a Fourier series with transverse (F) and longitudinal (Z) coefficients [3-5].

The transverse magnetization after n pulses and n dephasing periods with phase increment θ can therefore be represented as

$$M_+(z) = \sum_{n=-N}^{n=N} F_n e^{inz\theta}.$$ (3.16)

The longitudinal magnetization M_z after the nth pulse is given by

$$M_z(z) = \sum_{n=-N}^{n=N} Z_n e^{inz\theta(TR)}.$$ (3.17)

Note that $Z_n = Z_n^*$, since Z_n is always real and that the longitudinal "twists" are sinusoids.

We can therefore define the transverse magnetization phase twists as sub-states defined as follows:

$$F_n = \int_0^1 M_{xy}(z) e^{-inz\theta} dz,$$ (3.18)

$$F_{-n}^* = \int_0^1 M_{xy}^*(z) e^{-inz\theta} dz,$$ (3.19)

and the longitudinal magnetization sinusoids as sub-states defined as follows

$$Z_n = \int_0^1 M_z(z) e^{-inz\theta(t)} dz.$$ (3.20)

These sub-states can be easily propagated in MR sequences according to the following transition rules [3-6] from the pre-pulse state (negative subscript) to the post-pulse state (positive subscript).

1. Gradients increase F_n, i.e., dephasing: $F_n^+ \rightarrow F_{n+1}^-$, or decrease F_{-n}^{*+}, i.e., rephasing: $F_{-n}^{*+} \rightarrow F_{-n+1}^{*-}$.

2. RF pulses mix sub-states between F_n, F_{-n}, and Z_n as given in the transition matrix:

$$\begin{bmatrix} F_n \\ F_{-n}^* \\ Z_n \end{bmatrix}^+ = \begin{bmatrix} \cos^2\left(\dfrac{\alpha}{2}\right) & e^{2i\phi}\sin^2\left(\dfrac{\alpha}{2}\right) & -ie^{i\phi}\sin(\alpha) \\ e^{-2i\phi}\sin^2\left(\dfrac{\alpha}{2}\right) & \cos^2\left(\dfrac{\alpha}{2}\right) & ie^{-i\phi}\sin(\alpha) \\ \dfrac{-i}{2}e^{-i\phi}\sin(\alpha) & \dfrac{i}{2}e^{i\phi}\sin(\alpha) & \cos(\alpha) \end{bmatrix} \begin{bmatrix} F_n \\ F_{-n}^* \\ Z_n \end{bmatrix}^-.$$ (3.21)

3. T_2 relaxation attenuates F_n, i.e., $F_n^+ \rightarrow F_{n+1}^- = E_2 F_n^+$, where $E_2 = e^{-\frac{t}{T_2}}$.

4. T_1 relaxation attenuates Z_n, i.e., $Z_n^+ \rightarrow Z_n^- = E_1 Z_n^+$, for $n \neq 0$, where $E_1 = e^{-\frac{t}{T_1}}$.

5. T_1 relaxation results in recovery of Z_0^+, i.e., $Z_0^+ \to Z_0^- = E_1 Z_0^+ + M_0\left(1-E_1\right)$.

Let us consider a simple pulse sequence comprising of two RF pulses, applied around the x-axis ($\phi = 0$). The initial magnetization is at equilibrium, i.e., $M_0 = M_z = Z_0 = 1$. The magnetization after the first excitation pulse (α_1) will be $F_0^+ = -i\sin(\alpha_1)$, $F_0^{*+} = i\sin(\alpha_1)$, and $Z_0^+ = \cos(\alpha_1)$. This corresponds to a transverse magnetization (M_{xy}) of amplitude $\sin(\alpha_1)$, aligned along the $-y$-axis and a remaining longitudinal magnetization (M_z) of amplitude $\cos(\alpha_1)$. During the time period $t = \tau$, F_0^+ will evolve into $E_2 F_1^-$ and F_0^{*+} will evolve into $E_2 F_1^{*-}$. Applying the transition matrix for the second RF pulse (α_2) will result in two fully dephased states F_1^+ and F_{-1}^{*+}:

$$F_1^+ = iE_2 \cos^2\left(\frac{\alpha_2}{2}\right)\sin(\alpha_1), \tag{3.22}$$

$$F_{-1}^{*+} = E_2 \sin^2\left(\frac{\alpha_2}{2}\right)\sin(\alpha_1). \tag{3.23}$$

Since F_{-1}^{*+} has a reversed phase history during the next time period τ, the F_{-1}^{*+} state will refocus to form a spin echo (F_0), at time $t = 2\tau$, of amplitude $E_2 \sin^2\left(\frac{\alpha_2}{2}\right)\sin(\alpha_1)$, while the F_1^+ state will continue to dephase generating the F_2^- sub-state. As well as producing the two dephased transverse sub-states the second RF pulse will also create a further F_o sub-state, that will continue to dephase, and an additional longitudinal state (Z_1).

If $\alpha_1 = \frac{\pi}{2}$ and $\alpha_2 = \pi$, then $F_0 = 1.0 \times E_2$, i.e., the classic 90°–180° spin echo. However, if $\alpha_1 = \alpha_2 = \frac{\pi}{2} = 90°$, then $F_0 = 0.5E_2$. This is the classic Hahn 8-ball echo described previously.

3.2.1   EXTENDED PHASE GRAPH

A simple way to represent these various states is to use an extended phase graph (EPG), where the dephasing of the transverse magnetization is plotted against time [3-7]. Between RF pulses the phase is a straight line with a slope proportional to the precession rate. The EPG essentially depicts magnetization in spatial Fourier space, with the calculation of the echo amplitudes obtained through the above matrix multiplications. The EPG for a 90°–180° spin echo is shown in Figure 3.7 while the EPG for two α 90° pulses is shown in Figure 3.8.

The EPG for a gradient echo sequence comprising of a train of identical α pulses is shown in Figure 3.9. Note that the third echo is combination of the primary free-induction decay from the third α pulse together with the rephased F_{-1}^{*+} magnetization from the second α pulse.

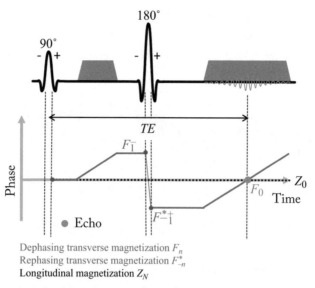

Figure 3.7: Simple extended phase graph for a 90°–180° spin echo with gradient dephasing/rephasing.

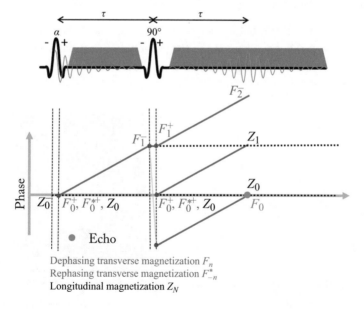

Figure 3.8: EPG for a 90°-90° Hahn echo with gradient dephasing/rephasing.

Additional spoiler gradients may also be included to ensure complete dephasing across a voxel. If we continue our EPG simulation for 65 α pulses and we look at the amplitude of the echoes (F_o state) we see strong oscillations at the beginning of the pulse train that converge to a steady state after about 40 pulses (Figure 3.10a). Figure 3.10b shows the generation and flow of the transverse states over the first 65 RF pulses. There is a linear increase in the number of pathways. Note that the term "gradient spoiling" is a misnomer, gradients cannot destroy transverse magnetization they can only dephase it; only effects such as RF pulse rotations, relaxation or diffusion can destroy magnetization.

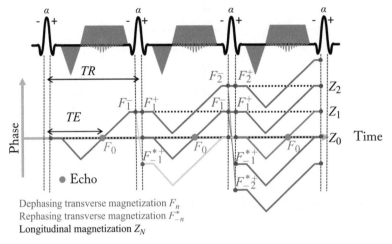

Figure 3.9: EPG for a train of four α pulses each comprising a gradient recalled echo sequence. Note the third echo (pink pathway) is a combination of the primary free-induction decay from the third α pulse together with the rephased F_{-1}^{*+} magnetization from the second α pulse (yellow pathway).

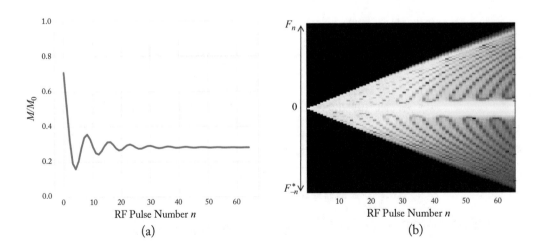

Figure 3.10: (a) Amplitudes of the echo (F_0) for the first 65 α pulses for a gradient-spoiled gradient echo. (b) The full extended phase graph showing the amplitude of the transverse magnetization states.

RF-spoiling involves the use of a train of RF pulses with a specific phase increment scheme that effectively destroys the residual transverse magnetization. The phase of the RF excitation pulse (ϕ_j) is incremented, usually quadratically, on each excitation according to the following scheme:

$$\phi_j = \tfrac{1}{2}\phi_0\,(j^2 + j + 1), j = 0,1,2,\dots \tag{3.24}$$

There are certain "magic numbers" for the for the phase increment (ϕ_0) that can produce excellent spoiling across a range of T_1, T_2 and excitation flip angles [3-8] and is equal to the steady-state transverse magnetization for a perfectly spoiled sequence, i.e., no T_2 contribution, as given by the Ernst equation for a flip angle α:

$$M_{ss} = M_0 \frac{\left(1 - e^{-\frac{TR}{T_1}}\right)}{\left(1 - \cos\left(\alpha\right)\cdot e^{-\frac{TR}{T_1}}\right)} \cdot \sin\left(\alpha\right). \tag{3.25}$$

Figure 3.11 shows the calculated signal as a function of quadratic phase increment. Note the good agreement between the simulation and the Ernst signal at $\phi_0 = 117°$. This value works well across a range of T_1 and T_2 values. Figure 3.12 shows the EPGs for three RF-spoiled gradient echo sequences with different phase increments. A vlaue of 117° is commonly used in commercial MR systems.

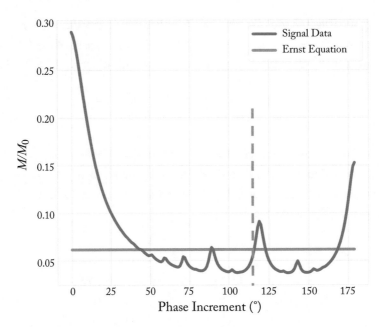

Figure 3.11: Signal in a RF-spoiled gradient echo sequence as a function of quadratic phase increment. The Ernst signal for perfect spoiling is also shown (red line). Note that both signals match at a phase increment of 117°. The simulated parameters were: T_1 = 112 ms, T_2 = 97 ms, TE = 3.3 ms, and TR = 7.1 ms.

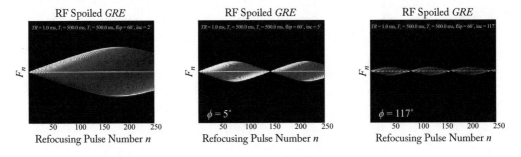

Figure 3.12: This figure shows RF-spoiled EPGs for three different RF spoiling increments (ϕ_0). Note how the spoiling creates a sinusoidal oscillation of spreading and confluence with the oscillation frequency increasing with increased phase increment demonstrating the cancellation of the steady-state transverse component.

As mentioned before *FSE/TSE* sequences can be performed with reduced and/or variable refocusing flip angles. For example, the use of refocusing pulses < 180° give rise to a number of coherence pathways which lead to the creation of a so-called pseudo-steady-state (*PSS*) [3-9]. Figure 3.13a shows the echo amplitudes for an echo train using 90° refocusing pulses. Note the steady-state amplitude of 0.71, which is approximately 30% less than the amplitude for a sequence using 180° refocusing pulses, but with effectively a 75% reduction in power deposition (since power deposition is proportional to the square of the flip angle). Also note the initial signal modulation during the initial echoes. Figure 3.13b shows the EPG demonstrating the evolution of the transverse sub-states.

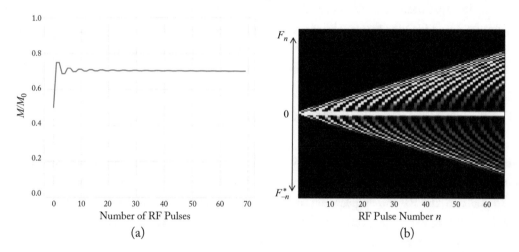

(a)   (b)

Figure 3.13: (a) Echo amplitudes during a 70 pulse CPMG sequence, with a flip angle of 90° and ignoring relaxation effects. Note the signal modulation during the initial echoes. (b) Extended phase graph (EPG) for a 70 pulse CPMG sequence using a refocusing flip angle of 90°, showing the evolution of the transverse substates. The color scale is proportional to the magnitude of F_n. Relaxation effects are ignored in this example.

3.3   CONVENTIONAL METHODS FOR RELAXOMETRY

Here we will discuss conventional MRI methods for the quantification of relaxation times, as well as introduce new methods for generating additional contrasts from a single-pulse sequence.

3.3.1   INVERSION RECOVERY ESTIMATION OF T_1

The gold standard method for measuring T_1 relaxation is the multiple inversion recovery (IR) method, discussed in Section 3.1.4. In the IR method the initial longitudinal magnetization M_0 is inverted by the application of an RF pulse that nutates it by 180°. The z-component of the magnetization (M_z) then recovers with time (t) via T_1 relaxation as described by the Bloch equation that assumes that T_1 relaxation follows first order rate kinetics, i.e., an exponential recovery,

$$\frac{dM_z(t)}{dt} = \frac{M_0 - M_z(t)}{T_1}.$$

(3.26)

After a selected inversion time (TI), M_z is nutated into the transverse plane so it can be measured. In an IR-based imaging sequence this usually involves a spin echo sequence to sample the nutated M_z with the shortest possible echo time (*TE*) to minimize any transverse dephasing, i.e., T_2 effects. Since M_z has been nutated into the transverse plane it is necessary to then wait enough time, i.e., the repetition time (*TR*) needs to be long enough, for M_z to fully recover back to thermal equilibrium, i.e., M_0. Once this has happened the whole experiment can be repeated using a different inversion time. In this way the T_1 recovery curve can be sampled at several T_1 values and the T_1 can be estimated by fitting the following equation to the data

$$S\left(TI\right) = S_0\left(1 - Ae^{-\frac{TI}{T_1}}\right) \tag{3.27}$$

where $S(TI)$ is the measured signal intensity at each T_1 value and S_0 is the signal that, theoretically, could be obtained from M_0. Effectively, S_0 can be considered to represent the relative proton density (ρ). In addition, the RF flip angles need to be exact and uniform across the region being sampled. Errors can be due to a variety of factors including mis-calibration of the pulse amplitude and poor slice profile. These effects are most commonly addressed using adiabatic inversion pulses, such as the hyperbolic secant, that will perfectly invert the magnetization for a B_1 field above a certain threshold. This pulse also gives a uniform excitation profile across the imaging slice. In the case of perfect inversion, the factor A, the inversion efficiency, is 2. However, in case of an imperfect inversion this parameter can also be fitted in addition to T_1 and S_0.

A complication of using the IR method is that when $TI < 0.693 \cdot T_1$, M_z is negative. Since it is usually the magnitude signal that gets reconstructed this negative sign information is lost, i.e., only the modulus is observed. While it is possible to fit the equation to the modulus data it is preferable, in order to reduce the variance in the fitted T_1 value, to restore the polarity of the data. There are several methods described in the literature to achieve this result. Figure 3.14a shows the above equation fitted to (a) the magnitude data and (b) the polarity restored data obtained at 11 different inversion times ranging from 50–4,000 ms. An advantage of using modulus data in clinical IR-prepared imaging is that by judicious selection of TI it is possible to null the signal from a particular tissue. For example, in Figure 3.14a the signal is nulled at approximately 300ms. In clinical MR neuroimaging the TI value is often set to null the signal from cerebrospinal fluid, which typically has a very high signal. Such a sequence is known as FLuid Attenuated Inversion Recovery (FLAIR).

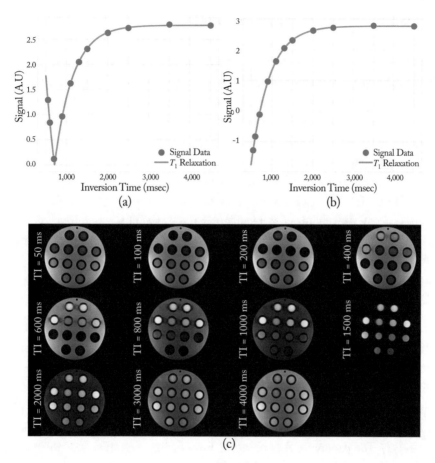

Figure 3.14: Inversion recovery data acquired at eleven different inversion times ranging from 50–4,000 ms. The T_1 of this sample is estimated to be 438 ms, S_0 is estimated to be 2,776 and the inversion efficiency is 1.6: (a) shows the magnitude data where the negative magnetization for the first three inversion times is shown as the absolute value; (b) shows the situation where the polarity of the first three data points in corrected; and (c) shows the magnitude images of a test object containing gels with different relaxation times at each TI.

The IR method ideally requires $TR > 5 \cdot T_1$, so that M_z is fully recovered before the next experiment. This also has the advantage that any residual transverse magnetization will have decayed away before the next inversion pulse. However, this means that the use of a standard spin echo as the readout method is not particularly time efficient. Alternative readout strategies have been proposed including echo-planar imaging (EPI) where the entire image is acquired following a single inversion pulse. This has the advantage that the EPI readout does not affect the recovery of M_z. However, there may be image quality issues, particularly distortions, associated with the use of EPI readouts. Alternatively, fast or turbo spin echo readouts can be used. It is possible to acquire an entire image from a single inversion using half-Fourier single-shot fast spin echo (SSFSE) acquisitions or HASTE [3-2, 3-10, 3-11]. The main disadvantage of these types of acquisitions is the presence of T_2 decay during the echo train that can result in image blurring. Finally, gradient echo readouts can be used. However, since gradient echo sequences typically use reduced flip angles excitations, i.e., <90°, they affect the recovery of M_z, accelerating the relaxation and giving rise to incorrect T_1 values unless this effect is addressed in the calculation. In addition, if multiple gradient echo readouts are used after a single inversion pulse, M_z will be changing during the readout period potentially causing artifacts in the image.

There are numerous variants of the IR-prepared methods that use a variety of readout strategies in either 2D or 3D. The readouts are often segmented meaning that a fraction of the total number of lines of raw data required to reconstruct the image(s) are acquired following a single inversion pulse. The acquisitions are then repeated with several different TIs in order to estimate T_1.

3.3.2 FASTER ESTIMATION OF T_1 WITH INVERSION RECOVERY: LOOK-LOCKER METHODS

The Look-Locker (LL) methods differ from the IR methods described above in that they acquire data at multiple TIs from a single inversion pulse, i.e., during a single recovery period [3-12]. The readout method is typically a segmented, low flip angle, gradient echo. However, like the gradient echo readout described above the effect of the low flip angle pulses, is to accelerate the T_1 recovery. The result is that the observed T_1 relaxation time, often referred to as T_1^*, depends upon the TR of each gradient echo readout and the flip angle. To estimate a true T_1 a LL correction is applied where $T_1 = (A-1)T_1^*$. The parameter A is obtained from fitting Equation 3.27 to the data. The use of a balanced steady-state free precession (bSSFP) gradient echo readout perturbs the M_z recovery less and is the method of choice in a variant of the LL method used for myocardial T_1 mapping, commonly referred to as MOdified Look-Locker Imaging (MOLLI) [3-13]. The MOLLI method acquires several single-shot bSSFP acquisitions synchronized to the subject's ECG, interleaved with periods of recovery. The original MOLLI method implemented a 3(3)3(3)5 scheme, taking 17 heartbeats (HB), the numbers refer to the number of images acquired in subsequent HBs, while the numbers in brackets refer to the number of recovery HBs. The data acquisition scheme is shown in Figure 3.15. Other, potentially more efficient, schemes have also been proposed such as 5(3)3. One other notable alternative to MOLLI is the SAturation single-SHot Acquisition (SASHA) method that uses 90° saturation pulses rather than 180° inversion pulses and increments the time between the saturation pulse and the bSSFP readout in typically 10 heartbeats. The first image is acquired without a 90° pulse to provide an estimate of the fully relaxed magnetization.

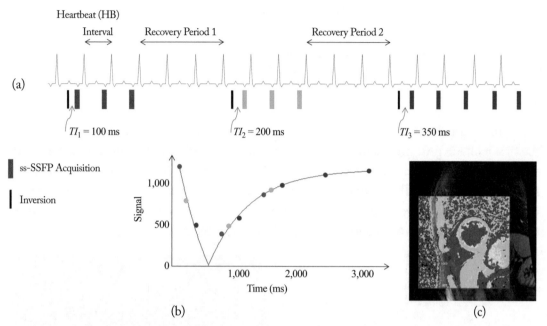

Figure 3.15: MOdified Look Locker Imaging (MOLLI): (a) illustrates the acquisition strategy for the classic MOLLI sequence that requires 17 heartbeats (HB). Each image acquisition is based upon a single-shot balanced SSFP sequence (ss-SSFP) acquired at end-diastole. The sequence employs three 180° inversion pulses. Following the first trigger the first image is acquired with a *TI* of 100 ms, the subsequent image is acquired with a *TI* (TI_1) of 100 ms + 1 × HB interval and the third image with a *TI* of 100 ms + 2 × HB intervals (red points). There is then a magnetization recovery period of 3 HB intervals, followed by a further three acquisitions but this time with the initial *TI* (TI_2) set to 200 ms (green points). After three further acquisitions and a second recovery period there is a final set of images acquired with an initial T_1^* (TI_3) of 350 ms (blue points); (b) shows the relationship of the acquisitions to the T_1 recovery curve. A magnitude inversion recovery model is fitted to the data points to determine the T_1; and (c) Shows a calculated myocardial T_1 map.

3.3.3 ESTIMATIONS OF T_1 WITH VARIABLE FLIP ANGLE GRADIENT ECHO

An alternative method to estimate T_1 is to use a short *TR*, RF-spoiled, gradient echo sequence with different flip angles. The advantage of an RF-spoiled gradient echo sequence is that even in the case where $TR \ll T_2$ there is no residual transverse magnetization. This means that the signal is a function of T_1, *TR* and the flip angle (α) if the *TE* is kept very short. T_1 can therefore be calculated by acquiring several RF-spoiled gradient echo images with different flip angles and fitting the following equation, known as the Ernst equation, to the data

$$S(\alpha) = S_0 \left[\frac{\left(1 - e^{-\frac{TR}{T_1}}\right)\sin(\alpha)}{1 - \cos(\alpha)e^{-\frac{TR}{T_1}}} \right]. \tag{3.28}$$

This method is most commonly used with 3D RF-spoiled gradient echo sequences in which the slice profile is perfectly rectangular. The method is still sensitive to RF pulse calibration errors, so it is advisable to map the B_1 transmit field variation across the field-of-view using an appropriate B_1-mapping technique. The resulting maps can then be used to spatially correct the estimated T_1 values. Figure 3.16 shows the above equation fitted to data obtained using a 3D spoiled gradient echo sequence excitation flip angles of 2°, 5°, 12°, 17°, 22°, and 27°.

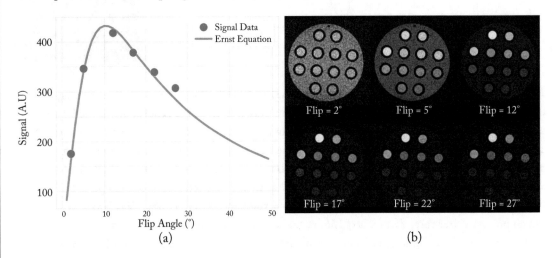

(a)   (b)

Figure 3.16: Data obtained using a 3D spoiled gradient echo sequence with a *TE* of 1.9 ms and a *TR* of 5.1 ms and excitation flip angles of 2°, 5°, 12°, 17°, 22°, and 27°. (a) In this example, the T_1 was estimated to be 314 ms and S_0 estimated to be 4,860, and ((b) shows the images of a test object containing gels with different relaxation times at each flip angle.

3.3.4   T₂ ESTIMATIONS WITH MULTIPLE-ECHO SPIN ECHO

The gold-standard for measuring T_2 relaxation is the multiple-echo spin echo (MESE) method. The initial longitudinal magnetization M_0 is tipped into the transverse plane by the application of a 90° RF pulse, where it dephases due to T_2 relaxation. Following a time τ after the 90° excitation pulse a 180° pulse is used to flip the dephased transverse magnetization about the axis of the pulse. This phase reversal means that the previously dephasing magnetization precesses in the opposite sense and naturally rephases producing a spin echo signal a further time τ later, i.e., 2τ after the 90° pulse. This time is referred to as the echo time (*TE*). The effect of the refocusing pulse is to effectively reverse the dephasing due to static field non-uniformities. However, the dephasing due to intrinsic T_2 relaxation is irreversible and the transverse magnetization (M_{xy}) will decay with time as described by the Bloch equation. Again, assuming first-order rate kinetics the T_2 relaxation can be expressed as follows:

$$\frac{dM_{xy}(t)}{dt} = -\frac{M_{xy}(t)}{T_2}.$$

(3.29)

It is therefore possible to quantify T_2 relaxation by forming echoes at different *TE*s by increasing the time τ between the 90° excitation pulse and the refocusing pulse. However, since a long *TR* is required to allow full M_z recovery before the next excitation this is very time consuming. In 1954, Carr and Purcell showed that a spin echo sequence using sequential refocusing pulses could create a train of sequential echoes, i.e., at increasing *TE*s, from the same excitation pulse. This combination of pulses can be written as $90°_x - \tau - [180°_x - 2\tau]_n$ and is known as a Carr-Purcell sequence. The subscript x denotes the axis along which the pulse is applied in the transverse plane. This method has the advantage that as well as reducing the time to acquire multiple echoes, it also dramatically reduces the effect of molecular self-diffusion on the determination of T_2, since diffusion only occurs during the period 2τ. In this way the T_2 decay curve can be sampled at several *TE* values and the T_2 can be estimated by fitting the following equation to the data

$$S(TE) = S_0 e^{-\frac{TE}{T_2}},$$

(3.30)

where $S(TE)$ is the measured signal intensity at each *TE* value and S_0 can be considered to represent the relative proton density (ρ). This equation only applies if the *TR* is sufficiently long $TR > 5 \cdot T_1$ to ensure that that M_z is fully recovered before the next experiment.

As with T_1 mapping any miscalibration of the RF pulse amplitude can result in erroneous measurements of T_2. This problem was identified in the early days of NMR and can be addressed using an appropriate RF phase cycling scheme. The preferred embodiment is the so-called Carr–Purcell–Meiboom–Gill (CPMG) method, $[90°_x] - \tau - [180°_y - 2\tau]_n$, where n is the number of echoes.

Within this method the excitation and refocusing pulses are applied orthogonally to each other. This simple modification makes the accuracy of the 180° refocusing pulse less critical. However, when using slice-selective refocusing pulses the required flip angle falls off toward the edges of the slice. Ideally the refocusing pulses should be non-selective, often known as "hard" pulses, but this is obviously incompatible with interleaved multi-slice T_2 mapping. A potential compromise is to make the slice width of the refocusing pulse double that of the excitation pulse and use an appropriate spacing between the multiple slices. It should also be noted that any spins that do not receive a perfect 180° refocusing will ultimately generate stimulated echoes (see Section 3.1.3) in a long train of echoes, that may introduce a T_1 weighting into the echo train resulting in erroneous T_2 values. Figure 3.17 shows the above equation fitted to 16 echoes of a CPMG acquisition.

3.3.5   MULTIPLE-ECHO FAST/TURBO SPIN ECHO

Given the relatively long acquisition time to perform quantitative T_2 measurements, several approaches have been used to try and reduce the acquisition time. It is possible to modify a fast or turbo spin echo (FSE/TSE) sequence, in which the echoes in an CPMG echo train are individually phase encoded to reduce the overall acquisition time, to create a fast, multi-echo. acquisition. For example, an FSE sequence with a total echo train length (ETL) of 16 could divide the acquisition into 4 different echo images, each comprised of 4 phase-encoded echoes.

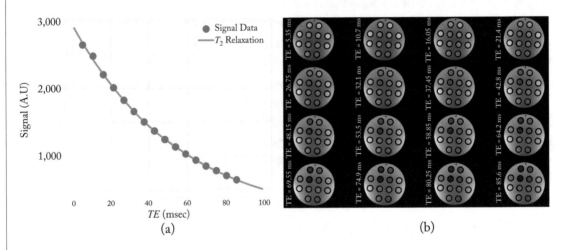

Figure 3.17: (a) Shows data acquired at 16 echoes of a CPMG acquisition with *TE*s at a multiple of 5.35 ms and a *TR* of 2500 ms. The T_2 was estimated to be 55.8 ms and S_0 is estimated to be 2,962. (b) Shows the images of a test object containing gels with different relaxation times at each *TE*.

3.3.6 T_2 ESTIMATIONS WITH T_2 PREPARATION

An alternative to multi-echo readouts is to encode the T_2 weighting into a preparation scheme that can be applied prior to a time efficient readout scheme. For example, quantitative myocardial T_2 mapping has been performed with a T_2-prepared single-shot SSFP readout The T_2 preparation [3-14] consists of a Malcolm Levitt (MLEV) sequence $[90°_x] - \tau - [180°_x] - 2\tau - [180°_x] - 2\tau - [-180°_x] - 2\tau - [-180°_x] - \tau - [-90°_x]$, where $\tau = TE/8$.

The 180° pulses are composite pulses $\{[90°_x] - [180°_y] - [90°_x]\}$ and the final -90° "tip-up" pulse is also composite $\{[270°_x] - [-360°_x]\}$ to provide more uniform off-resonance behavior. Such a preparation scheme can also be applied to volumetric (3D) sequences as well. The T_2 is calculated from a small number of different TE values.

3.3.7 T_2^* ESTIMATIONS WITH MULTIPLE TEs

As mentioned above, T_2^* relaxation results primarily from non-uniformities in the static magnetic field. These non-uniformities may be due to limitations in the achievable homogeneity of the MRI system magnet itself or from susceptibility-induced distortions produced by the subject being imaged. The former is minimized by the process of "shimming" the magnet, where small pieces of steel (known as "shims") are appropriately positioned inside the bore of the magnet to compensate for the inherent design and manufacturing limitations. The latter is an inevitable consequence of the magnetism associated with tissue or other materials placed inside the magnetic field. There are several diseases that can affect T_2^* relaxation such as iron overload in the heart and/or liver. Quantitation of the T_2^* relaxation time can be used as a biomarker of disease progression and to quantify the tissue iron concentration.

The gold-standard for measuring T_2^* relaxation is a multi-echo gradient echo sequence. Since a gradient echo forms the echo signal through a gradient reversal rather than an RF refocusing pulse the contribution due to static field non-uniformities is not eliminated and the T_2^* decay curve can be sampled at several TE values and the T_2^* can be estimated by fitting the following equation to the data

$$S(TE) = S_0 e^{-\frac{TE}{T_2^*}}. \tag{3.31}$$

Multiple echoes are generally achieved by reversing the polarity of the frequency encoding gradient. Due to the difference in precessional frequencies between hydrogen nuclei in water and fat the signal from a multi-echo gradient echo can also be modulated by their periodic coming in- and out-of-phase. Care should therefore be taken in considering which echoes to use in the estimation of tissue T_2^* relaxation times.

3.4   MULTI-PARAMETRIC QUANTITATIVE MRI (mqMRI)

There has always been an interest in reducing the acquisition time of MRI—while maximizing the number of image contrasts. New methods can acquire additional quantitative maps efficiently within a single acquisition.

New multi-parametric quantitative mqMRI methods have included: multi-step or interleaved (ZTE), SSFP (Fingerprinting), and FSE contrast based (MAGIC).

3.5   INTERLEAVED CONTRASTS

The simplest method to obtain multiple contrasts with a single scan is to interleave more than one type of contrast generation strategy within a single sequence [3-15–3-17]. Two examples of this are shown in Figure 3.18, which obtains both T_2^* and T_2 contrast by using a T_2 preparation with multiple variable TEs (τ in Figure 3.18a), and reading out at multiple readout time points (TE_r). MASE [3-16] and SAGE-EPI [3-17] acquire additional GRE data between the 90° and inversion pulses. These methods become complicated for quantification due to inaccuracies of slice profiles, signal refocusing during T_2^* decay, and T_2 and T_2^* decay during each readout [3-16]. NEATR-SMS [3-18] combines deep learning with SAGE-EPI to obtain parametric maps.

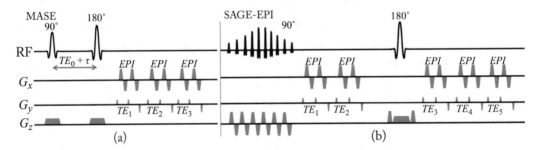

Figure 3.18: A straightforward method for generating multiple types of contrasts within a single sequence is to interleave the contrasts available, as is done with (a) MASE and (b) SAGE, where data with different T_2 and T_2^* contrasts are obtained.

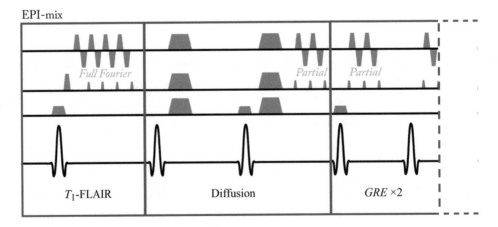

Figure 3.19: EPI-mix: An EPI method that uses several preparation modules to interleave T_1, T_2 and diffusion contrast. The acquisition consists of fully sampled and undersampled EPI datasets that are combined to generate maps. One possible ordering of contrasts and EPI acquisitions is shown here, with GRE repeated multiple times in the third block.

EPI-mix [3-19] is another EPI method with mixed contrast (Figure 3.19). EPI-mix incorporates multiple "modules" that are repeated throughout its acquisition to obtain multiple contrasts. Each module might be put in different orders, and acquire either partial or fully sampled k-space data. The modules incorporate T_1-FLAIR, T_2-FLAIR, diffusion, and GRE contrasts. Portions of each module might be combined to generate additional contrasts. The data can be used to fit signal models to generate quantitative maps.

3.5.1   3D RADIAL WITH ZERO ECHO TIME

Zero-echo time (ZTE) MRI involves a radial sequence that begins its k-space trajectory during the RF excitation resulting in the center of k-space being missed. This data is separately acquired or estimated. ZTE mqMRI (also known as "3D Silent Data Mining") [3-20] involves two steps: a T_1-dominated step after an inversion pulse and a T_2-dominated step after T_2 preparation (Figure 3.18). After each preparation, the readout is repeated during a train of pulses (each with flip angle α). The signal model can then be fitted to the data in order to obtain quantitative maps. The train of pulses follows the equation

$$M_{z,n} = M_{z,0} E_1^n \cos^n \alpha + M_0 \left(1 - E_1\right)\left(1 - E_1^n \cos^n \alpha\right) / \left(1 - E_1 \cos \alpha\right). \tag{3.32}$$

This model assumes that $M_{z,n}$ after the nth inversion pulse is perfect, and is therefore $M_{z,n+} = -M_{z,n-}$. Similarly, it assumes that after T_2 preparation, the signal is attenuated by T_2, such that $M_{z,n+} = -M_{z,n-} e^{-TE/T_2}$. The model is then fit by least-squares minimization to obtain values for T_1 and T_2.

3D Silent Data Mining

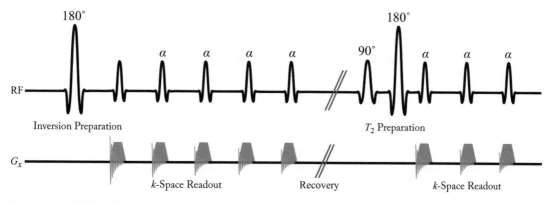

Figure 3.20: "Silent data mining" is a zero echo time sequence that acquires T_1 weighted followed by T_2 weighted data that can be fit to generate quantitative maps.

ZTE mqMRI has low (so considered to be essentially silent) acoustic noise due to the radial acquisition. The low flip angles used will generate a native proton density (ρ) contrast with minimal

T_1 saturation [3-21, 3-22]. 3D ZTE mqMRI may be of interest for lung imaging, although it has not yet been demonstrated for that purpose.

3.6   MULTI-PARAMETRIC SSFP

3.6.1   STEADY-STATE METHODS

As previously discussed, SSFP can create T_1 and T_2 contrast, which enables the generation of multiple contrasts simultaneously. The small flip angle, steady-state signal approximation of SSFP [3-23] is (where E_1 and E_2 are given in Equation 3.21)

$$S_{steady-state} = S_0 \frac{\left(1-E_1\right)\sin\alpha}{1-\left(E_1-E_2\right)\cos\alpha-E_1 E_2}. \tag{3.33}$$

An inversion pulse before the SSFP sequence generates additional T_1 contrast [3-24, 3-25], which then follows an equation related to the time after the pulse—described by the nth repetition of the TR:

$$S(nTR) = S_{ss}\left[1-S_0/S_{ss}\cdot\exp\left(\frac{-nTR}{T_1^*}\right)\right]. \tag{3.34}$$

T_1^* in this equation is a combination of both T_1 and T_2,

$$T_1^* = \left[1/T_1\cos^2\left(\alpha/2\right)+1/T_2\sin^2\left(\alpha/2\right)\right]^{-1}. \tag{3.35}$$

The signal is then fitted to determine T_1 and T_2.

Linear least squares fitting is a fast computational method for calculating the intercept, a, and slope, b, from a vector of known parameters, x, and measured parameters, y,

$$y = a + x \times b. \tag{3.36}$$

The DESPOT1 method [3-26] is a common method for obtaining quantitative relaxation values by linearizing Equation (3.35), but using a spoiled gradient echo sequence such that E_2 is zero. DESPOT1 linearisation results in an equation of the form

$$\frac{S(\alpha)}{\sin\alpha} = M_0\left(1-E_1\right)+\frac{S(\alpha)}{\tan\alpha}E_1, \tag{3.37}$$

where S is the signal obtained from measurement, E_1 is equal to exp($-TR/T_1$), and α can either assumed from the pulse sequence setting or as a measured parameter. For this linearization, the intercept is $a = M_0(1-E_1)$ and the slope is $b = E_1$. After fitting for the slope, b, a T_1 measurement can be obtained:

$$T_1 = -TR/\ln[b].$$ (3.38)

Using a linear least squares (LLS) fit, the vectorized calculation of the slope from the **x** and **y** vectors is

$$b = \frac{N\Sigma(x_i y_i) - \Sigma x_i \Sigma y_i}{N\Sigma(x_i^2) - (\Sigma x_i)^2}.$$ (3.39)

More advanced fitting methods can be used, which are discussed in Chapter 5.

DESPOT2 is similar to DESPOT1 [3-26], although it uses SSFP, where E_2 is not null in Equation (3.35). The linearized form of DESPOT2 is

$$\frac{S(\alpha)}{\sin\alpha} = M_0(1-E_1)\frac{E_2}{1-E_1 E_2} + \frac{S(\alpha)}{\tan\alpha} \times (E_1 - E_2)/(1 - E_1 E_2),$$ (3.40)

where, in this case, $E_2 = \exp(-TR/T_2)$. DESPOT2 is related to DESPOT1, although with more parameters, with the intercept $a = M_0 (1 - E_1) \frac{E_2}{1 - E_1 E_2}$ and the slope $b = (E_1 - E_2)/(1 - E_1 E_2)$. DESPOT2 uses knowledge of T_1 from a prior fit (such as from a DESPOT1 measurement) to obtain the T_2 values from the fitted slope. T_2 is then equal to

$$T_2 = -TR/\ln\left[\frac{b-E_1}{bE_1 - 1}\right].$$ (3.41)

Both DESPOT1 and DESPOT2 are prone to errors from flip angle estimation inaccuracies, and a flip angle map is often used to reduce these errors. DESPOT2 is also affected by any additional inaccuracies from T_1 estimations.

3.6.2 TRANSIENT-STATE METHODS

Transient state imaging involves acquiring data whilst the magnetization is evolving from prior RF pulses, i.e., the magnetization is not in a steady-state. MR Fingerprinting (MRF) [3-27] is a form of transient-state imaging that uses inversion recovery preparation followed by a variable flip angle SSFP (IR-vSSFP) readout to generate multiple image contrasts (see Figure 3.21). The variable flip angles follow a pseudo-random pattern. MR Fingerprinting then uses a Bloch simulation signal variation, often using the EPG formalism as discussed above, to generate a "dictionary." The dictionary is a look-up table of possible signal patterns for different possible T_1 and T_2 values. The acquired signal patterns and generated signal patterns are "matched" to find the optimum fit. MRF, and other transient state methods, can be performed with either fully sampled or undersampled *k*-space. The latter is commonly used to increase acquisition speed. Other parameters might be incorporated and randomized into both the acquisition and simulation, such as B_1, B_0, *TR*, or *TE*. Transient-state contrast decoding is discussed more fully in Chapter 5.

MRF gets its name from the random looking transient signal evolutions, which appear like "fingerprints," and are unique for different T_1s and T_2s. The relative proton density (ρ) is the scaling factor used to match the simulated signal evolution with the measured signal.

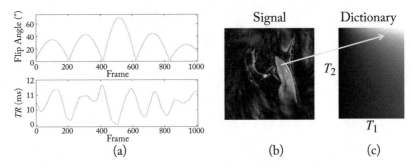

Figure 3.21: (a) A pseudorandom variation of flip angles and *TR* that is commonly used for MRF repeatability studies. The *TR* is sometimes held fixed rather than randomized. (b) An image at one time point within the 979 frames. Undersampling artifact can be seen, as well as the multiple contrasts between tissue. (c) A dictionary, for the same time point as in (a), showing the magnitude of the signal for different possible T_1 and T_2 values.

MR-STAT [3-28] (Magnetic Resonance Spin TomogrAphy in Time-domain) is similar to MRF, as it uses a pseudorandom flip angles in the steady state. MR-STAT reconstruction occurs within the *k*-space domain, rather than in the image domain as used in the MRF reconstruction method.

3.7   MULTI-PARAMETRIC FSE: syMRI/MAGIC

3.7.1   syMRI/MAGIC T$_1$

SyMRI, also known as MAGnetic resonance Image Compilation (MAGIC) [3-29], uses the sequence "QRAP-MASTER" [3-30, 3-31] ("Quantification of Relaxation times And Proton density by Multiecho Acquisition of a Saturation Recovery using TSE Readout"). QRAP-MASTER begins its sequence with a saturation/inversion pulse before a turbo/fast spin echo readout, which generates two forms of contrast from the saturation and the excitation pulses (θ and α, respectively; see Figure 3.22). QRAP-MASTER is then repeated with variations in these two angles or with pulse timing differences.

The QRAP-MASTER analytical solution can be found by separating the equation into two parts (and ignoring the 180^0 pulses) along the T_1 recovery curve, both equations are very similar:

$$M_{TD} = M_0 - \left(M_0 - M_{TR}\cos\theta\right)e^{-TD/T_1}, \tag{3.42}$$

and

$$M_{TR} = M_0 - \left(M_0 - M_{TD}\cos\alpha\right)e^{-(TR-TD)/T_1}, \tag{3.43}$$

where θ is the initial saturation pulse, which is commonly 120° or 180°; α is the excitation pulse, which rotates the longitudinal magnetization into the transverse direction for read-out; TR is the time between two θ pulses; and, TD is the time between the θ and α pulses. Note that TD is the delay time and could also be replaced with TI, or inversion time, with only a minor change of meaning.

These two equations are then combined to find the steady-state solution, which then gives the signal equation of QRAP-MASTER [3-30, 3-31]:

$$S_{TD} = S_0 \left[\frac{1 - \left(1 - \cos\theta\right)e^{-TD/T_1} - \cos\theta e^{-TR/T_1}}{1 - \cos\theta\cos\alpha e^{-TR/T_1}} \right] - \left(S_0 - S_{TR}\cos\theta\right)e^{-TD/T_1}. \tag{3.44}$$

This equation enables T_1 and M_0 (= S_0) estimation through signal fitting. The M_0 calculation is much more easily performed at the T_2 estimation stage, as described in the next section.

QRAP-MASTER (*MAGIC, SyMRI*)

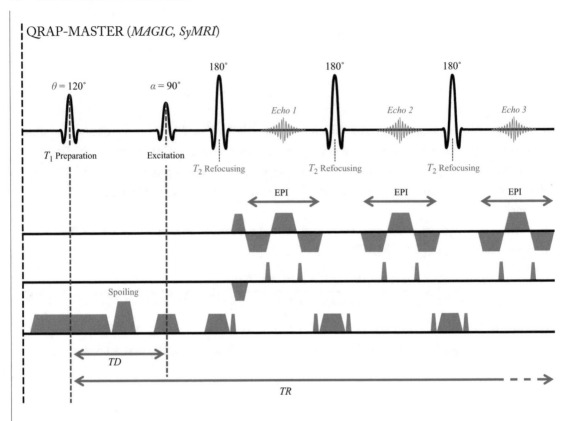

Figure 3.22: The QRAP-MASTER sequence uses a slice selective saturation pulse for T_1 preparation (= θ, often 120° or 180°), followed by an excitation pulse (= α, often 90°). A spin-echo acquisition accelerated with EPI follows these combined preparations. The acquisition is then repeated with multiple θ and α angles, which can then be fit to the equations in Section 3.7.

3.7.2   syMRI/MAGIC T_2

The above equation describes the signal effects of T_1, but not T_2. The T_2 refocusing pulses cause a standard exponential decay after slice selection. Very simply, the magnetization is

$$S_{TE} = S(0)e^{-TE/T_2} . \tag{3.45}$$

The starting magnetization, $S(TE{=}0)$, is the magnetization immediately after the α pulse, which is attenuated by $\sin(\alpha)$ from the rotation of the longitudinal magnetization into the transverse plane. This signal is *also* attenuated by the flip angle, α, caused by the coil sensitivity profile. In order to obtain a *relative* proton density (ρ) measurement, the signal $S(0)$ that results from T_2 exponential fitting should be scaled by these, such that the ρ is

$$\rho = \frac{S(0)}{\alpha \sin(\alpha)}.$$

$$(3.46)$$

This relative proton density is a proton density measurement that is affected by arbitrary scaling factors, such as those that the scanner introduces for digital processing. The inclusion of all factors would include additional receiver hardware, field strength, temperature, voxel size, and sample loading effects.

3.7.3 syMRI/MAGIC B_1

An B_1 effective field map is found from the ratio of the magnetization before and after the saturation pulse, θ. The signal then depends only on the flip angle, θ:

$$\theta_{eff} = \cos^{-1}\left(\frac{M_{T0}}{M_{TR}}\right).$$

$$(3.47)$$

This could be applied similarly for direct measurement of the flip angle α, although the slice selection complicates the estimation.

3.8   CONCLUSION

In this chapter, we discussed acquisition methods for the quantification of T_1, T_2, and T_2^*. We discussed both conventional and fast quantification MRI methods, although further discussion of advanced spatial and contrast reconstructions follows in the next chapters.

BIBLIOGRAPHY

[3-1]   E. L. Hahn, Spin echoes, *Phys. Rev.*, 80(4), pp. 580–594, Nov. 1950. DOI: 10.1103/Phys-Rev.80.580. 41

[3-2]   J. Hennig, A. Nauerth, and H. Friedburg, RARE imaging: A fast imaging method for clinical MR, *Magn. Reson. Med.*, 3(6), pp. 823–833, Dec. 1986. DOI: 10.1002/mrm.1910030602. 47, 60

[3-3]   D. E. Woessner, Effects of diffusion in nuclear magnetic resonance spin-echo experiments, *J. Chem. Phys.*, 34(6), pp. 2057–2061, Jun. 1961. DOI: 10.1063/1.1731821. 50

[3-4]   J. Hennig, Echoes—how to generate, recognize, use or avoid them in MR-imaging sequences. Part I: Fundamental and not so fundamental properties of spin echoes, *Concepts Magn. Reson.*, 3(3), pp. 125–143, Jul. 1991. DOI: 10.1002/cmr.1820030302. 50

[3-5]   M. Weigel, Extended phase graphs: Dephasing, RF pulses, and echoes—pure and simple, *J. Magn. Reson. Imaging*, 41(2), pp. 266–295, Feb. 2015. DOI: 10.1002/jmri.24619. 50

[3-6]   K. Scheffler, A pictorial description of steady-states in rapid magnetic resonance imaging, *Concepts Magn. Reson.*, 11(5), pp. 291–304, Jan. 1999. DOI: 10.1002/(SICI)1099-0534(1999)11:5<291::AID-CMR2>3.0.CO;2-J. 51

[3-7]   J. Hennig, Multiecho imaging sequences with low refocusing flip angles, *J. Magn. Reson.*, 78(3), pp. 397–407, Jul. 1988. DOI: 10.1016/0022-2364(88)90128-X. 52

[3-8]   Y. Zur, M. L. Wood, and L. J. Neuringer, Spoiling of transverse magnetization in steady-state sequences, *Magn. Reson. Med.*, 21(2), pp. 251–263, Oct. 1991. DOI: 10.1002/mrm.1910210210. 55

[3-9]   D. C. Alsop, The sensitivity of low flip angle RARE imaging, *Magn. Reson. Med.*, 37(2), pp. 176–184, Feb. 1997. DOI: 10.1002/mrm.1910370206. 56

[3-10]   R. F. Busse, S. J. Riederer, J. G. Fletcher, A. E. Bharucha, and K. R. Brandt, Interactive fast spin-echo imaging, *Magn. Reson. Med.*, 44(3), pp. 339–348, Sep. 2000. DOI: 10.1002/1522-2594(200009)44:3<339::AID-MRM1>3.3.CO;2-E. 60

[3-11]   R. C. Semelka, N. L. Kelekis, D. Thomasson, M. A. Brown, and G. A. Laub, HASTE MR imaging: Description of technique and preliminary results in the abdomen, *J. Magn. Reson. Imaging*, 6(4), pp. 698–699, Jul. 1996. DOI: 10.1002/jmri.1880060420. 60

[3-12]   M. Karlsson and B. Nordell., Analysis of the Look-Locker T1 mapping sequence in dynamic contrast uptake studies: simulation and in vivo validation, *Magn. Reson. Imaging*, 18(8), pp. 947–954, 2000. DOI: 10.1016/S0730-725X(00)00193-4. 60

[3-13]   D. R. Messroghli, A. Radjenovic, S. Kozerke, D. M. Higgins, M. U. Sivananthan, and J. P. Ridgway, Modified Look-Locker inversion recovery (MOLLI) for high-resolutionT1 mapping of the heart, *Magn. Reson. Med.*, 52(1), pp. 141–146, Jul. 2004. DOI: 10.1002/mrm.20110. 60

[3-14]   P. Kellman, A. H. Aletras, C. Mancini, E. R. McVeigh, and A. E. Arai, T2-prepared SSFP improves diagnostic confidence in edema imaging in acute myocardial infarction compared to turbo spin echo, *Magn. Reson. Med.*, 57(5), p. 891, May 2007. DOI: 10.1002/mrm.21215. 64

[3-15]   R. D. Newbould et al., Perfusion mapping with multiecho multishot parallel imaging EPI, *Magn. Reson. Med.*, 58(1), pp. 70–81, Jul. 2007. DOI: 10.1002/mrm.21255. 66

[3-16]   C. Wang et al., Simultaneous dynamic R2, R2', and R2* measurement using periodic π pulse shifting multiecho asymmetric spin echo sequence moving estimation strategy: A feasibility study for lower extremity muscle, *Magn. Reson. Med.*, 77(2), pp. 766–773, Feb. 2017. DOI: 10.1002/mrm.26126. 66

[3-17]   H. Schmiedeskamp et al., Combined spin- and gradient-echo perfusion-weighted imaging, *Magn. Reson. Med.*, 68(1), pp. 30–40, Jul. 2012. DOI: 10.1002/mrm.23195. 66, 65

[3-18]   B. Bilgic et al., Highly accelerated multishot EPI through synergistic machine learning and joint reconstruction, *arXiv Prepr.*, vol. arXiv:1808, Aug. 2018. 66

[3-19]   S. Skare et al., A 1-minute full brain MR exam using a multicontrast EPI sequence, *Magn. Reson. Med.*, 79(6), pp. 3045–3054, Jun. 2018. DOI: 10.1002/mrm.26974. 67

[3-20]   A. B. Wiesinger, F. Janich, M. Ljungberg, E. Barker, and G. Solana, 3D MR parameter mapping using magnetization prepared zero TE," in *Proc. Intl. Soc. Mag. Reson. Med.*, 2018, p. 0061. 67

[3-21]   M. Weiger, K. P. Pruessmann, and F. Hennel, MRI with zero echo time: hard versus sweep pulse excitation, *Magn. Reson. Med.*, 66(2), pp. 379–389, 2011. DOI: 10.1002/mrm.22799. 68

[3-22]   D. M. Grodzki, P. M. Jakob, and B. Heismann, Ultrashort echo time imaging using pointwise encoding time reduction with radial acquisition (PETRA), *Magn. Reson. Med.*, 67(2), pp. 510–518, Feb. 2012. DOI: 10.1002/mrm.23017. 68

[3-23]   K. Sekihara, Steady-state magnetizations in rapid NMR imaging using small flip angles and short repetition intervals, *IEEE Trans. Med. Imag.*, 6(2), pp. 157–164, 1987. DOI: 10.1109/TMI.1987.4307816. 68

[3-24]   P. Schmitt et al., Inversion recovery TrueFISP: Quantification of T1, T2, and spin density, *Magn. Reson. Med.*, 51(4), pp. 661–667, Apr. 2004. DOI: 10.1002/mrm.20058. 68

[3-25]   K. Scheffler, On the transient phase of balanced SSFP sequences, *Magn. Reson. Med.*, 49(4), pp. 781–783, Apr. 2003. DOI: 10.1002/mrm.10421. 68

[3-26]   S. C. L. Deoni, B. K. Rutt, and T. M. Peters, Rapid combined T1 and T2 mapping using gradient recalled acquisition in the steady state, *Magn. Reson. Med.*, 49(3), pp. 515–526, Mar. 2003. DOI: 10.1002/mrm.10407. 68, 69

[3-27]   D. Ma et al., Magnetic resonance fingerprinting, *Nature*, 495(7440), pp. 187–192, Mar. 2013. DOI: 10.1038/nature11971. 69

[3-28]   A. Sbrizzi et al., Fast quantitative MRI as a nonlinear tomography problem, *Magn. Reson. Imaging*, 46, pp. 56–63, Feb. 2018. DOI: 10.1016/j.mri.2017.10.015. 70

[3-29]   L. N. Tanenbaum et al., Synthetic MRI for clinical neuroimaging: Results of the magnetic resonance image compilation (MAGiC) prospective, multicenter, multireader trial, *Am. Soc. Neuroradiol.*, 38(6), pp. 1103–1110, 2017. DOI: 10.3174/ajnr.A5227. 70

[3-30]   J. B. M. Warntjes, O. D. Leinhard, J. West, and P. Lundberg, Rapid magnetic resonance quantification on the brain: Optimization for clinical usage, *Magn. Reson. Med.*, 60(2), pp. 320–329, Aug. 2008. DOI: 10.1002/mrm.21635. 70, 71

[3-31]   J. B. M. Warntjes, O. Dahlqvist, and P. Lundberg, Novel method for rapid, simultaneous T1, T2*, and proton density quantification, *Magn. Reson. Med.*, 57(3), pp. 528–537, Mar. 2007. DOI: 10.1002/mrm.21165. 70, 71

CHAPTER 4

Spatial Decoding

Magnetic resonance signals are encoded in *k*-space using gradient pulses. Image reconstruction is the process of decoding these signals. In Chapter 2, the basics of continuous Fourier transformations were reviewed, where signals in *k*-space were related to spatial locations in an image. Here, we review the theory behind reconstructing images from digital, discrete MRI signals, deriving the effect of non-Cartesian acquisitions. In order to produce images on a Cartesian grid, non-uniform *k*-space acquisitions must be re-gridded. In addition, fast acquisitions that do not sample a complete *k*-space produce imperfect point spread functions, and produce aliasing in the final images. Strategies involving multiple receiver coils and nonlinear reconstructions, commonly used to anti-alias accelerated acquisitions will be discussed here.

4.1 THE DISCRETE FOURIER TRANSFORM

Analog signals in MRI are sampled by data acquisition boards with finite sampling bandwidths in order to be converted to digital signals. The maximum allowable image resolution is then defined by this bandwidth and follows the Nyquist-Shannon theorem. This theorem states that a band-limited signal with bandwidth, *B*, can be completely reconstructed from its samples if they are sampled at a rate no larger than $\frac{1}{2}B$, hence defining the achievable image resolution for a *fully sampled* experiment of discrete points. The final image resolution is usually the highest achievable by the Nyquist/Shannon theorem. This criterion is commonly referred to as the *Nyquist criterion*.

Fourier operations are written in a discrete form for image reconstruction. The forward and inverse Fourier transform on a discrete grid of *N* Cartesian points can be written as:

$$\text{Forward: } S(k) = \sum_{n=0}^{N-1} s(n) e^{-j2\pi nk/N}$$

$$\text{Inverse: } s(n) = \sum_{k=0}^{N-1} S(k) e^{j2\pi nk/N} .$$

(4.1)

Since these are linear operations, an $N \times N$ transformation can also be written in a matrix form as $\boldsymbol{S} = \boldsymbol{Fs}$, where

$$F = \frac{1}{\sqrt{N}} \begin{bmatrix} 1 & 1 & 1 & \cdots & 1 \\ 1 & w & w^2 & \cdots & w^{N-1} \\ 1 & w^2 & w^4 & \ddots & w^{2(N-1)} \\ \vdots & \vdots & \vdots & \ddots & \vdots \\ 1 & w^{N-1} & w^{2(N-1)} & \cdots & w^{(N-1)(N-1)} \end{bmatrix} , \quad (4.2)$$

where $w = e^{-j2\pi/N}$, and $\boldsymbol{F}\,\boldsymbol{F^H} = \boldsymbol{F^H}\,\boldsymbol{F} = \boldsymbol{I}$.

4.2   NON-CARTESIAN k-SPACE RECONSTRUCTION

Although many MRI experiments sample data on a discrete uniformly spaced Cartesian grid, as seen in Chapter 3, non-Cartesian data sampling patterns (e.g., spiral and radial) have unique advantages, such as their robustness to undersampling. To reconstruct images from radial k-space, projection algorithms can be used as well as gridding approaches. Filtered back projection (FBP) was originally developed for CT as an approximate solution to the inversion of the Radon transform [4-1]. FBP uses the central slice theorem where the radial spokes acquired in k-space represent the Fourier transform of the image projections, and the operations performed are similar to those employed in computed tomography (CT). Although FBP is still widely used in nuclear medicine, FBP has been replaced in MRI by gridding techniques that interpolate data onto a 2D regular grid. *Regridding* or simply *gridding* aims at performing a transformation from a non-uniform grid in the frequency domain into a uniformly-sampled image domain grid. Uniformity is not necessarily a requirement of the image domain, but methods achieving images on non-uniform grids are beyond the scope of this text. Following Jackson et al. [4-2], if one considers a two-dimensional image I(x,y), its Fourier transform is given by:

$$c\left(k_x, k_y\right) = \int_{-\infty}^{+\infty} I\left(x, y\right) e^{-ik_x \cdot x} \cdot e^{-ik_y \cdot y} \cdot dx \; dy . \quad (4.3)$$

If the index l going from 1 to L is associated with the discrete samples acquired, the k-space sampling trajectory $t(k_x, k_y)$ is approximated by a series of 2D delta functions at positions k_x (l) and k_y (l):

$$t\left(k_x, k_y\right) = \sum_{l=1}^{L} \delta\left[k_x - k_x\left(l\right), k_y - k_y\left(l\right)\right]. \quad (4.4)$$

The sampled signal is given by:

$$c_s\left(k_x, k_y\right) = c\left(k_x, k_y\right) \cdot t\left(k_x, k_y\right). \quad (4.5)$$

In gridding, the sampled data are interpolated onto a Cartesian grid after convolving with a suitable interpolation weighting function, $f(k_x, k_y)$, often called a *convolution kernel*, and re-sampling onto a unit-spaced grid (sampled, convolved and sampled - *scs*):

$$c_{scs}\left(k_x,k_y\right)=\left[c_s\left(k_x,k_y\right)*f\left(k_x,k_y\right)\right]\cdot III\left(k_x,k_y\right)=\left\{\left[c\cdot t\right]*f\right\}\cdot III,\qquad(4.6)$$

where "*" is the symbol for convolution and $III(k_x, k_y)$ is the Dirac comb function, which consists of the sum of equispaced 2D delta functions that represent a 2D Cartesian grid. To obtain the effect on the ideal image ($I(x,y)$) of sampling, convolving, and sampling in *k*-space, I_{scs}, we apply a Fourier transformation to Equation (4.6):

$$I_{scs}\left(x,y\right)=\left\{\left[I\left(x,y\right)*T\left(x,y\right)\right]\cdot F\left(x,y\right)\right\}*III\left(x,y\right),\qquad(4.7)$$

where $T(x, y)$ and $F(x, y)$ are the inverse Fourier transforms of $t(k_x, k_y)$ and $f(k_x, k_y)$, respectively. Ignoring $T(x, y)$, we can see that the result of the gridding is multiplied by the inverse Fourier transform of the convolution kernel.

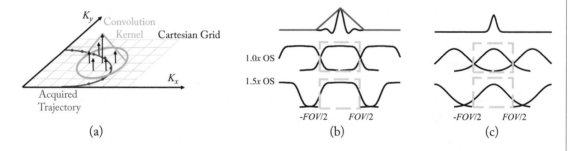

Figure 4.1: (a) Gridding kernel convolution and resampling; (b) effect of oversampling on a windowed SINC gridding kernel; and (c) effect of oversampling on a single-lobe kernel.

To obtain a perfect rectangular FOV, the ideal convolution function $f(k_x, k_y)$ would be a SINC [=sin(x)/x] function with a full-width [at] half-maximum (FWHM) of one pixel. However, the SINC function is defined over infinitely many points, which is impossible to implement in practice. For practical reasons, the SINC convolution is windowed to a local region (see Figure 4.1a). The use of a windowed SINC produces classical image aliasing from adjacent samples as well as an increase in central image intensity, or "apodization." These aspects can be traded off using simpler, single lobe kernels (see Figure 4.1b,c). De-apodization can also be achieved by dividing the resulting image by $F(x, y)$. Aliased replicas in the image domain can be pushed out of the *FOV* by using an oversampled *k*-space grid, although this increases computational complexity, which can be an issue for large matrix sizes.

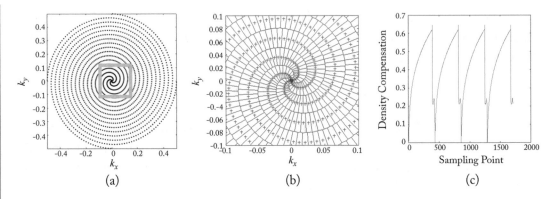

Figure 4.2: Non-Cartesian sampling patterns have, by definition, a non-uniform k-space density. This non-uniformity results in regions of the sampled k-space with a high and low sampling densities, which can result in image artifacts. For this reason, density correction is applied prior to convolution to correct for non-uniform sampling of the k-space. For radial or projection reconstruction, the density compensation factors can be derived analytically, while other trajectories require numerical estimations to establish a k-space mesh, such as a Voronoi diagram. The trajectory in (a), for instance, is a 4-arm spiral trajectory, while in (b) the mesh associated with the Voronoi diagram can be observed in the central part of the k-space. The curves in (c) represent the density compensation factors (DCFs) associated with the points in the trajectories, proportional to the area elements derived in (b), hence to the inverse of the area density function, or the density of samples per unit area.

In addition to the gridding kernel, another important factor is the convolution with the inverse Fourier transform of the trajectory sampling function ($[I(x, y) * T(x, y)]$). This can be addressed by adding a density compensation weighting function, $w(k_x, k_y) = \frac{1}{t * f}$, representing the inverse of an area density function, or the density of k-space samples per unit area (see Figure 4.2). The sampled, weighted, convolved, and re-sampled (swcs) data can then be represented as:

$$c_{swcs}\left(k_x, k_y\right) = \left\{\left[c_s \cdot w\right] * f\right\} \cdot III = \left\{c \cdot \left[\frac{t}{t * f}\right] * f\right\} \cdot III. \tag{4.8}$$

While the corresponding image is:

$$I_{swcs}\left(x, y\right) = \left(\left\{I * \left[T^{*-1}\left(T \cdot F\right)\right]\right\} \cdot F\right) * III, \tag{4.9}$$

where $*^{-1}$ stands for deconvolution. The final operation to obtain gridding is given by de-apodization of the gridding kernel:

$$I^*\left(x, y\right) = \frac{I_{swcs}\left(x, y\right)}{F\left(x, y\right)}. \tag{4.10}$$

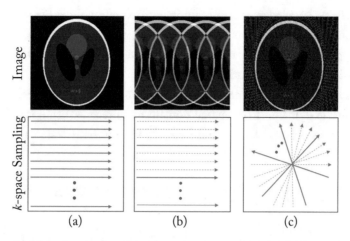

Figure 4.3: Simulated effect of undersampling on image reconstruction. Underdetermined systems create aliasing, dependent on k-space sampling: (a) fully sampled Shepp–Logan phantom; (b) Shepp–Logan phantom with ¼ of the k-lines acquired with a cartesian scheme (unacquired lines are dotted); and (c) Shepp–Logan phantom with ¼ of the k-lines acquired with a radial scheme.

As discussed above, the reconstruction of MR images from arbitrary k-space data is a linear operation going from coefficients in k-space to image pixels. When undersampled acquisitions are performed for speeding up examinations, the number of coefficients acquired in the k-space is smaller than the number of required image pixels. As the system of equations for linear reconstruction becomes underdetermined, multiple solutions exist for the linear reconstruction problem. As a result, if the missing data is simply zero-filled and reconstructed linearly, then aliasing will be present in the image (see Figure 4.3).

4.3   SPATIAL DECODING IN THE TRANSIENT STATE

Here, we discuss the effects of temporally changing trajectories, while sampling signals in the transient-state. Following the point spread function formalism from Stolk and Sbrizzi [4-3], we explore some of the implications of using transient-state acquisitions in combination with temporally-varying undersampled, non-Cartesian, k-space trajectories. We also consider other advanced reconstruction methods to further improve the results.

An image I_j consists of the effects from a perfect, noiseless signal, $S_j(x,y)$, and noise, $N_j(x,y)$:

$$I_j(x,y) = S_j(x,y) + N_j(x,y). \tag{4.11}$$

As seen in Chapter 3, by performing a simulation of the Bloch equations, it is possible to predict signal evolutions for a certain tissue with known physical properties. The vector of physical parameters, $\theta[T_1, T_2, \rho,..]$, includes, but is not limited to, relaxation times (T_1, T_2) and proton density (ρ). We consider that each location has a specific (but not-necessarily unique) vector of parameters, $\theta(x, y) = \theta[T_1, T_2, \rho,..]$. The impact of transient-state acquisitions, due to both contrast and spatial encodings, results in time-varying spatial effects.

The reconstruction of transient-state images involves operations going from the signal domain into the image domain. If the spatial domain to be reconstructed is a 2D image of N_x by N_y pixel indices, signals received from acquisitions in the transient-state can be written with an *encoding equation*:

$$s_{j,l} = \sum_{y=1}^{N_y} \sum_{x=1}^{N_x} M_j(\theta(x,y))\, e^{-ik_x(j,l)\cdot x}\, e^{-ik_y(j,l)\cdot y} + \sigma_{j,l}, \tag{4.12}$$

where M_j is the transverse magnetization at the j-th echo and $k_x(j,l)$ and $k_y(j,l)$ are the k-space sampling locations of the l-th readout sample during the j-th echo. $\sigma_{j,l}$ is complex Gaussian noise, which for a single receiver can be measured, and is modeled as a sample from a Gaussian distribution with zero mean.

After acquisition, signals are transformed into a stack of images associated with a temporal response. From these, we can reconstruct an image for every echo index, j, of the transient response, with a *decoding equation*:

$$I_j(x,y) = \frac{1}{N_x N_y} \sum_l w_{j,l}\, e^{ik_x(j,l)\cdot x}\, e^{ik_y(j,l)\cdot y}\, s_{j,l}, \tag{4.13}$$

where $w_{j,l}$ are a series of density compensation weights, to be applied if the sampling density is not uniform across k-space (see Figure 4.2). Ideally, the combination of the *encoding* and *decoding* steps would result in the intensity of each pixel of our image corresponding to the instantaneous transverse magnetization evolving following the Bloch equation, without additional point spread function (PSF) effects. However, in practice, especially when acquiring undersampled or non-Cartesian imaging, data will suffer from noise corruption and convolution with non-ideal PSFs. Therefore, we formulate the problem in general and assess the implications of each confounder separately.

Combining the *encoding and decoding* equations above, we can obtain for the signal:

$$S_j(x,y) = \frac{1}{N_x N_y} \sum_l \left(\sum_{x'=1}^{N_x} \sum_{y'=1}^{N_y} w_{j,l}\, M_j(\theta(x',y'))\, e^{ik_x(j,l)\cdot[x-x']}\, e^{ik_y(j,l)\cdot[y-y']} \right). \tag{4.14}$$

Following Stolk and Sbrizzi [4-3], we rewrite $S_j(x, y)$ as the convolution of $M_j(\theta(x,y))$ and a time-dependent point spread function, $P_j(x, y)$:

$$S_j(x,y) = \sum_{x'=1}^{N_x} \sum_{y'=1}^{N_y} M_j(\theta(x',y'))\cdot P_j(x-x',y-y'), \tag{4.15}$$

where $P_j(x, y) = \frac{1}{N_x N_y} \sum_l w_{j,l} e^{ik_x(j,l) \cdot x} e^{ik_y(j,l) \cdot y}$. The more similar the PSF, $P_j(x, y)$, is to a 2D Dirac delta function, δ, the more the final image pixel represent the underlying magnetization signal evolution.

Due to the linearity of convolutions, the decoded noise contribution, $N_j(x, y)$, is simpler to derive:

$$N_j(x,y) = \frac{1}{N_x N_y} \sum_l w_{j,l} \sigma_{j,l} e^{ik_x(j,l) \cdot x} e^{ik_y(j,l) \cdot y}. \tag{4.16}$$

4.4   FULL CARTESIAN SAMPLING

We now analyze the combination of encoding and decoding in several special cases, evaluating the effect of undersampling and non-Cartesian trajectories on image artifacts.

We start with the situation of "full" sampling, although full sampling in MRI is only an approximation as it is practically unachievable. By sampling in frequency space, resolving a single rectangular pixel would require infinite points in the frequency spectrum. Since we are limited to a finite set of coefficients in the Fourier domain, the image data will always be truncated, which is equivalent to a convolution with the Fourier transform of a *RECT* function, i.e., a *SINC*. So, full sampling in MRI is in theory impossible, but is commonly defined as having fulfilled the Nyquist criterion. If we want to reconstruct an image of X by Y pixels, we need to sample an equispaced grid in k-space of X by Y points.

4.4.1   CARTESIAN SAMPLING OF NOISELESS DATA

In order to compare signal and data models, a full Cartesian sampling of discrete points, without noise is the simplest case to consider. We start from the general equation in (4.14). In this case sampling is uniform, so $w_{j,l} = 1 \; \forall \{i, j\}$. With zero noise, $\sigma_{j,l} = 0 \; \forall \{i, j\}$:

$$I_j(x,y) = \frac{1}{N_x N_y} \sum_l \left(\sum_{x'=1}^{N_x} \sum_{y'=1}^{N_y} e^{ik_x(j,l) \cdot [x-x']} e^{ik_y(j,l) \cdot [y-y']} M_j \left(\theta(x',y') \right) \right). \tag{4.17}$$

The time-dependent point spread function is given by: $P_j(x, y) = \frac{1}{N_x N_y} \sum_l e^{ik_x(j,l) \cdot x} e^{ik_y(j,l) \cdot y}$. For the fully-sampled case, the index l is such that $k_x(j, l)$ and $k_y(j, l)$ sample the whole space within the Nyquist limits; and $P_j(x, y)$ is the Fourier transform of a *RECT*, or a 2D *SINC* function with a FWHM of one pixel, which is considered a good approximation to a two-dimensional Dirac δ. Leading to:

$$I_j(x,y) = \sum_{x'=1}^{N_x} \sum_{y'=1}^{N_y} SINC(x-x', y-y') M_j \left(\theta(x',y') \right). \tag{4.18}$$

The reconstructed image, $I_j(x, y)$, is the convolution between the real underlying data, M_j, and a *SINC* providing: $I_j(x, y) \cong M_j(\theta(x, y))$. This is the common assumption made when sampling with

MRI, where if the resolution is high enough the sidelobes of the *SINC* are assumed to be negligible, otherwise truncation artifacts (Gibbs ringing) may be seen.

For a 2D fully sampled Cartesian-encoding experiment in the absence of noise, the time evolution of the complex signal in each pixel can be directly compared with a transient-state simulation, making this an unbiased estimator.

4.4.2   CARTESIAN SAMPLING IN THE PRESENCE OF NOISE

In this case of fully sampled Cartesian data, with non-zero noise, $w_{j,l} = 1 \; \forall \{i,j\}, \; \sigma_{j,l} \neq 0$:

$$
\begin{aligned}
I_j(x,y) &= S_j(x,y) + N_j(x,y) \\
&= \frac{1}{N_x N_y} \sum_l \left(\sum_{x'=1}^{N_x} \sum_{y'=1}^{N_y} M_j\big(\theta(x',y')\big) \, e^{ik_x(j,l)\cdot[x-x']} \, e^{ik_y(j,l)\cdot[y-y']} \right). \\
&+ \frac{1}{N_x N_y} \sum_l \sigma_{j,l} \, e^{ik_x(j,l)\cdot x} \, e^{ik_y(j,l)\cdot y} \Big). \\
&= \sum_{x'=1}^{N_x} \sum_{y'=1}^{N_y} P_j(x-x', y-y') \, M_j\big(\theta(x',y')\big) + \frac{1}{N_x N_y} F(\sigma_{j,l}). \\
&= SINC(x,y) * M_j\big(\theta(x,y)\big) \, S_j(x,y) + \frac{1}{N_x N_y} F(\sigma_{j,l}).
\end{aligned}
\tag{4.19}
$$

In the presence of acquisition noise, for the fully sampled Cartesian case, the intensities in image space correspond to the evolution of the magnetization plus a zero-mean complex Gaussian noise.

4.5   FULL NON-CARTESIAN SAMPLING

4.5.1   TIME-DEPENDENT POINT-SPREAD FUNCTION

The difference between non-Cartesian sampling compared to Cartesian, is that the time-dependent PSF P_j will also include a density compensation factor different from 1, $w_{j,l}$:

$$
P_j(x,y) = \frac{1}{N_x N_y} \sum_l w_{j,l} \, e^{ik_x(j,l)\cdot x} \, e^{ik_y(j,l)\cdot y}.
\tag{4.20}
$$

When the data are fully sampled, $P_j(x, y)$ is a constant for all echoes j, and depends only on a single acquisition trajectory. With a fully sampled, non-Cartesian acquisition, the convolution function is not a *SINC* and depends on the sampling density across k-space. This induces a broadening of the point-spread function, leading to additional blurring of the final image. In addition, Equation (4.20) explains the robustness of common non-Cartesian k-space trajectories, such as radial or spiral, to undersampling. As the center of k-space is oversampled and the high frequencies are undersampled, $w_{j,l}$ is low at low frequencies and higher at high frequencies, hence aliasing from non-Cartesian acquisitions tends to be high frequency and appears more incoherent. An example of this behavior can be observed in Figure 4.3.

The noise statistics in the image are also influenced by w and will hence have a frequency dependency:

$$N_j(x,y) = \frac{1}{N_x N_y} \sum_l w_{j,l}\, \sigma_{j,l}\, e^{ik_x(j,l)\cdot x}\, e^{ik_y(j,l)\cdot y},$$ (4.21)

The effect of $w_{j,l}$ will be to amplify noise in k-space areas that are sampled less densely (usually the k-space edges, i.e., at the high spatial frequencies).

4.5.2   NON-CARTESIAN UNDERSAMPLING

We will now consider the cases where there are insufficient coefficients in k-space to reconstruct an image for each timeframe. Starting from more conventional undersampled images in the steady-state toward transient-state acquisitions.

Undersampling Noiseless Signals in the Steady State

In order to understand the effects of undersampling during temporal variations, it is easier to consider noiseless signals in the steady state. When the magnetization is in a steady-state, the acquisition may be considered a special case of the transient-state acquisition and $M_j(\theta(x,y))$ becomes independent of j. In this case, a set of undersampled images $I_j(x, y)$ are acquired after a sufficient number of RF excitation pulses bring the magnetization to a steady state, and then separate sections of k-space can be encoded. Normally, a single image is reconstructed from all time points that are combined because a single contrast is assumed over all available frames.

The point-spread function at each instant j is:

$$P_j(x,y) = \frac{1}{N_x N_y} \sum_l w_{j,l}\, e^{ik_x(j,l)\cdot x}\, e^{ik_y(j,l)\cdot y}.$$ (4.22)

Averaging over all values of j:

$$P(x,y) = \frac{1}{N_s} \sum_{j=1}^{N_x} P_j(x,y),$$ (4.23)

$$\hat{I}(x,y) = P * M(\theta(x,y)).$$ (4.24)

Unless $P(x, y)$ is a good approximation of a 2D Dirac δ, the final reconstruction will suffer from undersampling artifacts (Figure 4.4).

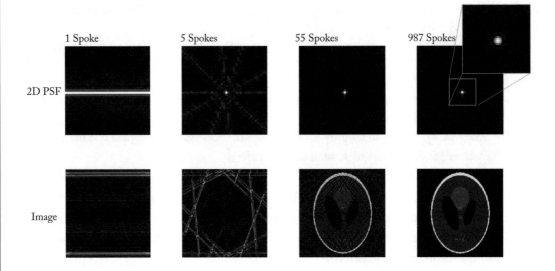

Figure 4.4: 2D PSF and reconstructed image of a Shepp-Logan phantom for, respectively: 1 radial spoke, 5 radial spokes, 55 radial spokes, and 987 radial spokes. Orientation of radial trajectories was incremented by the golden angle. When increasing the number of spokes, the PSF becomes gradually more localized, and striking artifacts due to aliasing are in turn reduced.

Undersampling During the Transient-State Evolution of the Magnetization

Let us assume that we are in a highly undersampled environment, and that the noise is negligible compared to undersampling artifacts. The general expression of the local signal in each pixel, where $S_j(x, y)$ is the signal contribution without noise is the combination of encoding and decoding, as seen in Equation (4.14) where the time-dependent point-spread function $P_j(x, y)$ was:

$$P_j(x,y) = \frac{1}{N_x N_y} \sum_l w_{j,l} \, e^{ik_x(j,l)\cdot x} \, e^{ik_y(j,l)\cdot y}. \tag{4.25}$$

If we call $P_j(x, y) = \frac{1}{N_s}\sum_{j=1} P_j(x, y)$ the PSF of the full sampling, with N_s equal to the total number of echoes in our acquisition, we can write an estimate of the fully-sampled case $\hat{S}_j(x, y)$, starting from the undersampled case, $S_j(x,y) = \sum_{x'=1}^{N_x} \sum_{y'=1}^{N_y} P_j * M_j(\theta(x',y'))$:

$$\hat{S}_j(x,y) = \sum_{x'=1}^{N_x} \sum_{y'=1}^{N_y} P_j * M_j(\theta(x',y')). \tag{4.26}$$

By applying $P = (P - P_j) + P_j$:

$$\hat{S}_j(x,y) = \sum_{x'=1}^{N_x} \sum_{y'=1}^{N_y} P_j * M_j\big(\theta(x',y')\big) + \sum_{x'=1}^{N_x} \sum_{y'=1}^{N_y} (P-P_j) * M_j\big(\theta(x',y')\big) = S_j(x,y) + \varepsilon_j(x,y),$$

(4.27)

where we have defined the *undersampling error* $\varepsilon_j(x, y) = \sum_{x'=1}^{N_x} \sum_{x'=1}^{N_x} (P - P_j) * M_j\big(\theta(x', y')\big)$, which depends on sampling trajectory as well as the evolution of the magnetizations $M_j\big(\theta(x, y)\big)$, hence also depends on the spatial distribution of the object.

4.6   ANTI-ALIASING

It is possible to resolve at least part of image aliasing by filling the missing datapoints in the image estimation model by using anti-aliasing strategies. The most common strategies used can be divided in *linear techniques* such as parallel imaging, or *nonlinear techniques* such as compressed sensing.

4.6.1   PARALLEL IMAGING

The spatial sensitivities of multiple RF receiver coils can be leveraged to reduce image aliasing caused by undersampling. Although the signal is highly correlated between adjacent receivers, each coil "sees" at least partially uncorrelated noise and has a unique sensitivity profile. Signal from different receiver coils can be utilized to resolve aliasing in undersampled acquisitions by solving for N_c sets of estimated images (where N_c is the number of coils), rather than a single set of equations. The most commonly used algorithms used for parallel imaging are GeneRalised Autocalibrating Partial Parallel Acquisition (GRAPPA) and SENSitivity Encoding (SENSE) [4-4]. For example, in SENSE (see Figure 4.5), individual coil images can be estimated as the multiplication of the object by the coil sensitivity map at that location:

$$F_1 = A_1 + B_1 = I_A C_{A1} + I_B C_{B1},$$

(4.28)

where F_1 is the aliased pixel for coil 1, C_{A1} and C_{B1} are the coil sensitivity for coil 1 at locations A and B, and I_A and I_B are the values of the pixels in the desired image at locations A and B (in Figure 4.5). In Equation (4.28), even if the values of the coil sensitivities C_{A1} and C_{B1} are known (for instance from a calibration step), there are two unknown values (the actual pixel values I_A and I_B), for a known F_1 value. We can write a similar equation for each receiver coil and pixel, allowing us to solve the system of equations for each location and unfold the aliased images. In contrast to SENSE, GRAPPA works directly in the undersampled k-space domain rather than in the aliased image domain.

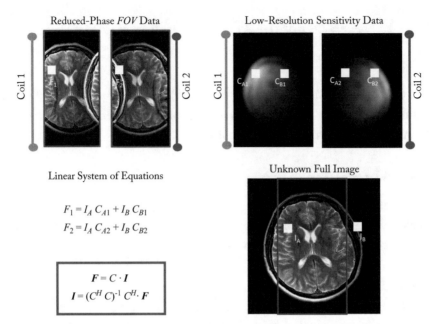

Figure 4.5: In SENSE folded coil images are unfolded utilizing coil sensitivities derived from fully sampled calibration. A linear model is used in order to recover spatial data in image domain.

The root-sum-of-squares (RSOS) algorithm is a common method to combine images from multiple coils, although this creates magnitude only data with a Rician biased noise-floor. The strength of the RSOS method is that it does not require knowledge of the individual coil sensitivities before combining the complex images from individual coils.

In some applications, estimates of the complex coil sensitivities are used in order to obtain also the phase of the combination of coil images. In order to optimally combine the complex information from each coil, the individual signals can be thought of as a stochastic signal process in the time domain $s(t)$ and an undesired noise process $n(t)$, also in the time domain. These signals are modeled in terms of their correlation statistics, and an optimization formula is used to compute an array filter maximizing the SNR for the assumed signal and noise statistics. The correlation matrices for temporal signal and noise statistics are given by:

$$R_s\left(j,k\right)=E\left[s_j\left(t\right)s_k\left(t\right)\right], \tag{4.29}$$

$$R_n\left(j,k\right)=E\left[n_j\left(t\right)n_k\left(t\right)\right], \tag{4.30}$$

for $j = 1\dots,N_{coils}$ and $k = 1\dots,N_{coils}$.

A filter maximizing the SNR power ratio is given by the eigenvector of the matrix $P = R_n^{-1} R_s$. For a proof, see [4-5, 4-6].

In order to apply this to the images acquired by individual receivers, the problem can be reformulated in the image domain as opposed to the time domain, an approach called *adaptive coil combination* [4-7]:

$$R_s(j,k) = \sum_{(x,y) \epsilon ROI} C_j(x,y)^* \, C_k(x,y), \qquad (4.31)$$

where sROI is a local set of coordinates around x and y, of arbitrary size. The noise correlation statistics can be estimated through a noise calibration scan or assumed to be white (i.e., R_n is the identity matrix). Under the assumption that R_n is the identity matrix, coil sensitivities can be estimated directly by computing the eigenvectors of R_s. In practice, this can be achieved by performing a singular value decomposition of R_s and taking the first right singular vectors.

4.6.2   COMPRESSED SENSING

In contrast to parallel imaging, which as formulated above uses linear algebra to combine different receivers, compressed sensing (CS) utilizes a nonlinear iterative reconstruction instead of a linear system of equations. Compressed sensing assumes that pixels are not independent, and the image is "compressible" in a given space. So, when data are transformed to a compressed space, a large number of coefficients will tend toward zero. A space with this characteristic is often called a "sparse" domain. There are many "sparse" domains that can be used, which depends on the properties of the data. A common example of a sparse representation is given by the discrete wavelet transform, at the basis of the compression algorithm adopted by JPEG. The sparse property of this transform can be readily demonstrated as JPEG compression can be applied to most medical images, reducing file size without noticing significant alterations (an example of compressed sensing using Wavelet regularization is seen in Figure 4.6).

There are three main requirements for compressed sensing.

- **Sparsifying transform:** There must be a sparse domain where most of the transformed coefficients tend toward zero.

- **Incoherent aliasing:** Undersampling must generate aliasing which is equally spread among datapoints in the sparse domain. Incoherent sampling will generate incoherent aliasing, which can be used to reduce the amount of data required.

- **Iterative reconstruction:** Nonlinear reconstruction is needed to remove the aliasing, so that consistency can be preserved with the acquired data.

Compressed sensing can be formulated as a constrained minimization problem. Iterative algorithms can be formulated if the forward and inverse transforms between signal and image domains are reversible, linear operations. If m represents the acquired data in the k-space and d is the resulting image, in a noiseless, fully sampled case then:

$$d = Fm \qquad m = F^{-1}d,$$

where F represents the linear transform between k-space and image space, while F^{-1} is its inverse. These are such that the direct application of F without using an iterative algorithm would produce an aliased image. As proposed by the seminal paper in compressed sensing MRI [4-8], the condition of sparsity, formulated theoretically as the condition of most coefficients being equal to zero in a sparse domain, is practically implemented as a minimization of the L_1 norm:

$$\underset{m}{\mathrm{argmin}} \ \| Fm - d \|^2 + \lambda \| \Psi m \|_1, \tag{4.32}$$

where ψ is the sparsifying transform. The L_2 term enforces the data consistency condition, while the L_1 enforces the sparsity condition. The factor λ is the regularization weighting parameter. One method to solve this is the projection onto convex sets (POCS) algorithm, also known as the *alternating projection* method [4-4-4-9]. An efficient approach to solve the minimization problem is also the *alternating direction method of multipliers* (ADMM), which solves the convex optimization problem by breaking it into smaller pieces, each of which have then better convergence properties [4-4-4-10].

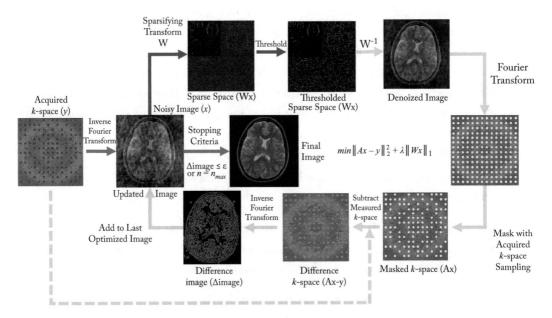

Figure 4.6: Compressed Sensing Flowchart, using a wavelet as a sparsifying transform. Some "typical" images of the iterative reconstruction are depicted for visualization purposes. Panel inspired by Blasche & Forman, "Compressed Sensing—the flowchart", from *MAGNETOM Flash Magazine*, 2017.

Although static images can be successfully recovered using compressed sensing, higher undersampling can be achieved when acquiring a temporal series. In dynamic imaging, the intrinsic correlation between frames can be exploited in the temporal domain because the frames share spatial features. An example of a sparsifying transformation for dynamic imaging is to perform finite differences in the time domain. Another example, in the case of periodic motion of, e.g., cardiac imaging, a sparsifying transform can be given by a pixel-wise Fourier transform of the time domain. Both of these approaches for sparsifying data are shown Figure 4.7.

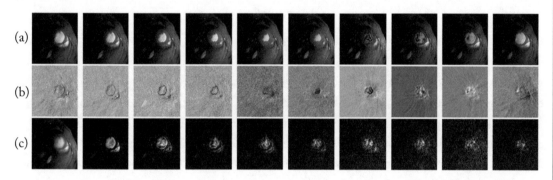

Figure 4.7: Examples of sparse representation of dynamic MRI: (a) CINE imaging of a mouse heart, showing the heart at different time-frames within the cardiac cycle; (b) the finite differences in the time domain between adjacent images in (a), showing that only a subset of pixels will change significantly between frames; and (c) the frequency components following a Fourier transformation of the time domain data (a), showing that the primary components have large contributions at low frequencies, while higher frequencies have small contributions.

4.6.3   LOW-RANK MODELS

Transient-state MRI as well as dynamic MRI data have some intrinsic correlations between time-frames. A low-rank approximation is a minimization problem where the cost function measures the fit between a given data matrix and a lower-dimensional low-rank matrix. If a low-rank domain exists, then the data can be approximated by a *subspace representation* [4-4–4-11]. Considering our complex temporal evolution, $M(\theta(x, y))$, which we can write as a vector $\boldsymbol{m} \in \mathbb{C}^{T \times N}$, with N coordinates (x and y pairs) and T timepoints. If there exists a transformation matrix $\boldsymbol{\Phi} \in \mathbb{C}^{T \times T}$ that is an orthonormal basis, such that $\boldsymbol{I} = \boldsymbol{\Phi}\boldsymbol{\Phi}^H$, then our vector can be rewritten $\boldsymbol{m} = \boldsymbol{\Phi}\boldsymbol{\Phi}^H \boldsymbol{m}$.

Considering a basis $\boldsymbol{\Phi} = [\varphi_1 \cdots \varphi_T]$, its h-dimensional subspace $\Phi_h = span\{\varphi_1 \cdots \varphi_h\}$ is a good approximation of the data if:

$$\|\boldsymbol{m} - \Phi_h \, \Phi_h^H \boldsymbol{m}\| < \epsilon, \tag{4.33}$$

where ϵ is the modeling error tolerance, and "H" indicates the Hermitian adjoint. If the norm in Equation (4.33) is the L_2 norm, a solution can be found by principal component analysis (PCA). A standard PCA method is SVD. If $d = Fm$ is the relation between data in the image domain d and data in the signal domain m:

$$
\begin{aligned}
d &= Fm \\
&= F\Phi\Phi\ m \\
&= F\Phi_h\Phi_h^H m \\
&:= F\Phi_h\alpha,
\end{aligned}
\qquad (4.34)
$$

where $\alpha = \Phi_h^H m$ are represented by the h temporal basis coefficients describing our data. If modeling error tolerance ϵ is neglected, we can also project the data back on the full signal domain $m = \Phi_h\ \alpha$.

A subspace representation of transient-state data can be obtained by performing a SVD on a set of simulated signal evolutions for certain combinations of T_1, T_2, etc. Data in any subspace has a reduced size, decreasing the memory needed for storage and reconstruction. Because the Fourier transformation is a linear operator and SVD compression is a linear summation, compression can be performed either before or after Fourier transformations. This linearity is especially advantageous because the number of computations required can be substantially reduced by performing the Fourier transformation operations after changing the data into a compressed subspace.

4.7   TRAJECTORY ERRORS

When gradient coils are rapidly switched, the varying magnetic field induces currents in other conducting surfaces, such as the magnet cryostat. These currents, known as "eddy currents," generate magnetic fields that distort the gradient waveforms and create artifacts in the images that are consistent with delays in the gradient waveforms. In addition to eddy currents, errors in hardware timing/calibration errors can also introduce delays between the received signals and the applied gradients [4-12].

For conventional Cartesian MRI, trajectory delays produce a shift of k-space in the readout direction, which results in a phase-shift of the reconstructed image, with no observable effect on the magnitude image. In contrast, when segmented or non-Cartesian k-space acquisitions are employed, significant artifacts are seen. A very common example where this effect is significant is echo planar imaging (EPI) (see Chapter 2) where each successive k-line is acquired in the opposed direction. Here, opposed k-space shifts in successive k-lines result in coherent ghosting artifacts (see Figure 4.8c). Similarly, in a non-Cartesian acquisition, shifted data will result in an erroneous assignment of k-space coordinates to the actual trajectory. However, in this case the effect will result in a more complex phase interference between acquisition interleaves rather than coherent ghosts. The effect of this trajectory error will be mostly visible as a low-frequency modulation of the image intensity. Some signal will "leak" from the object to areas around it, generating "shading" and "halos" (see Figure 4.8d).

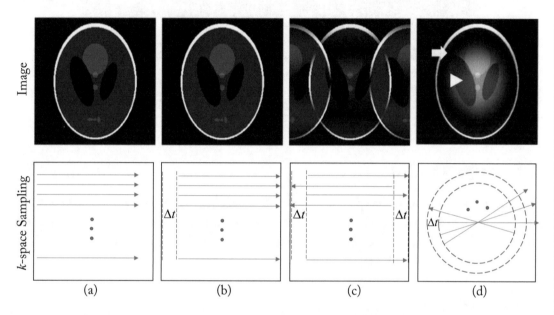

Figure 4.8: Simulated effect of a trajectory delay (1 pixel shift in k-space) on a numerical Shepp-Logan phantom: (a) no delay; (b) delay applied to standard Cartesian imaging, no effect is noticeable; and (c) delay on EPI acquisition, visible ghosting is present; and (d) the same delay in case of radial sampling, arrow points to shading, arrowhead to halos.

In some situations, eddy currents or errors in hardware calibration will produce not only gradient terms but static field (B_0) terms. These terms will produce a constant phase-shift along the readout. When MRI is acquired with standard Cartesian sampling, this phase addition is constant and will produce a static displacement of the whole image; in EPI or radial acquisitions different readouts will have multiple constant phase additions, resulting in phase interferences in k-space generating artifacts (see Figure 4.9).

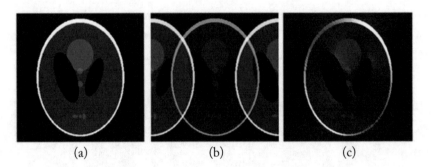

Figure 4.9: Simulated effects of B_0-induced phase errors (1 radians) for (a) Cartesian MRI, (b) EPI, and (c) Radial MRI.

4.8   CONCLUSION

This chapter demonstrated concepts associated with image reconstruction, compression and undersampling. In order to achieve quantitative imaging, contrast in the tissues of these images must now be decoded into quantifiable physical parameters. The next chapter reviews the basic concepts needed to perform contrast decoding.

BIBLIOGRAPHY

[4-1]   G. T. Herman, *Fundamentals of Computerized Tomography*. Springer Science & Business Media, 2009. DOI: 10.1007/978-1-84628-723-7. 78

[4-2]   J. I. Jackson, C. H. Meyer, D. G. Nishimura, and A. Macovski, Selection of a convolution function for fourier inversion using gridding, *IEEE Trans. Med. Imag.*, 10(3), pp. 473–478, 1991. DOI: 10.1109/42.97598. 78

[4-3]   C. C. Stolk and A. Sbrizzi, Understanding the combined effect of k-space undersampling and transient states excitation in MR Fingerprinting reconstructions, *IEEE Trans. Med. Imag.*, preprint 10.1109/TMI.2019.2900585, 2019. DOI: 10.1109/TMI.2019.2900585. 81, 82

[4-4]   A. Deshmane, V. Gulani, M. A. Griswold, and N. Seiberlich, Parallel MR imaging, *J. Magn. Reson. Imag.*, 36(1), pp. 55–72, 2012. DOI: 10.1002/jmri.23639. 87, 90, 91

[4-5]   S. M. Verbout, C. M. Netishen, and L. M. Novak, Polarimetric techniques for enhancing SAR imagery, *Synth. Aperture Radar*, 1630, pp. 141–173, 1992. DOI: 10.1117/12.59015. 88, 90, 91

[4-6]   P. B. Roemer, W. A. Edelstein, C. E. Hayes, S. P. Souza, and O. M. Mueller, The NMR phased array, *Magn. Reson. Med.*, 16(2), pp. 192–225, Nov. 1990. DOI: 10.1002/mrm.1910160203. 88, 90, 91

[4-7]   D. O. Walsh, A. F. Gmitro, and M. W. Marcellin, Adaptive reconstruction of phased array MR imagery, *Magn. Reson. Med.*, 43(5), pp. 682–690, May 2000. DOI: 10.1002/(SICI)1522-2594(200005)43:5<682::AID-MRM10>3.0.CO;2-G. 89, 90, 91

[4-8]   M. Lustig, D. Donoho, and J. M. Pauly, Sparse MRI: The application of compressed sensing for rapid MR imaging, *Magn. Reson. Med.*, 58(6), pp. 1182–1195, 2007. DOI: 10.1002/mrm.21391. 88, 90, 91

[4-9]   H. H. Bauschke and J. M. Borwein, On projection algorithms for solving convex feasibility problems, *SIAM Rev.*, 38(3), pp. 367–426, 2003. DOI: 10.1137/S0036144593251710. 90, 91

[4-10]   J. Assländer, M. A. Cloos, F. Knoll, D. K. Sodickson, J. Hennig, and R. Lattanzi, Low rank alternating direction method of multipliers reconstruction for MR fingerprinting, *Magn. Reson. Med.*, 79(1), pp. 83–96, 2018. DOI: 10.1002/mrm.26639. 90, 91

[4-11]   J. I. Tamir et al., T^2 shuffling: Sharp, multicontrast, volumetric fast spin-echo imaging, *Magn. Reson. Med.*, 77(1), pp. 180–195, Jan. 2017. DOI: 10.1002/mrm.26102. 91

[4-12]   G. Buonincontri, C. Methner, T. Krieg, T. A. Carpenter, and S. J. Sawiak, Trajectory correction for free-breathing radial cine MRI, *Magn. Reson. Imag.*, 32(7), pp. 961–964, 2014. DOI: 10.1016/j.mri.2014.04.006. 90, 92

CHAPTER 5

Contrast Decoding

In this chapter, we discuss the methods for parameter estimation by decoding the images acquired from the time-series data. In Chapter 3, we discussed the fundamentals of Bloch simulations and methods whereby multiple image contrasts could be obtained. Here, we focus our investigation on methods that can decode the contrast information in the steady-state or transient-state data, resulting in quantitative multi-parametric maps. We begin our formalism with the least squares approach, afterwards deriving a "pattern matching" algorithm based on a maximum inner product search used in magnetic resonance fingerprinting, thereafter exploring developments into compressed sensing anti-aliasing and machine learning routines.

5.1   INTRODUCTION

Magnetic resonance imaging time series can be used for local estimation of the parameters underlying a physical model of magnetization evolution. Here, we investigate methods to perform estimations of the signal by formulating the general inference problem from steady-state or transient-state data. We focus mostly on estimations of transient-state undersampled and non-Cartesian data as the general case; steady-state, Cartesian, or fully sampled data can be considered as a special case of this formalism.

5.2   LEAST SQUARES ESTIMATIONS

In Chapter 4, we defined the instantaneous, local magnetization $M_j\big(\theta(x, y)\big)$, where θ was a function of physical parameters, which included the location's proton density, T_1 and T_2, among any other parameters. We divide our temporal magnetization signal into a *static component*, or linear component $M_0(x, y)$, and a *dynamic component*, or nonlinear component $m(\vartheta(x, y),t_j)$, where $\vartheta(x, y)$ is a function of physical properties of the sample with the exclusion of the proton density. Thus, at echo number j and voxel location (x, y):

$$M_j\big(\theta\big(x,y\big)\big)=M_0\big(x,y\big)m\big(\vartheta\big(x,y\big),t_j\big). \tag{5.1}$$

We assume here that the equilibrium magnetization, $M_0(x, y)$, is static and unaffected by the temporal evolution of the magnetization. We also consider that $M_0(x, y)$ can be written as the multiplication of the local proton density $\rho(x, y)$, real-valued receiver sensitivity profile $B_1^-(x, y)$, and transceiver phase $e^{i\varphi(x,y)}$ (obtained as a combination of transmitter and receiver phases):

$$M_0(x,y) = \rho(x,y)B_1^-(x,y)e^{i\varphi(x,y)}. \tag{5.2}$$

With this formalism, we investigate the simultaneous estimation of M_0 and ϑ from transient-state data, starting with a least square formulation. Given the instantaneous image pixel, $d_j = I_j(x, y)$, it is possible to write a local estimation per each location (x, y) (thus dropping (x, y) from the equations):

$$\left(\widehat{M_0},\hat{\vartheta}\right) = \arg\min_{M_0,\vartheta} f\left(M_0,\vartheta\right) = \arg\min_{M_0,\vartheta} \sum_{j=1}^{N_t} \left| M_0 m\left(\vartheta, t_j\right) - d_j \right|^2, \tag{5.3}$$

which can be rewritten in L_2 −space vectors as:

$$\left(\widehat{M_0},\hat{\vartheta}\right) = \arg\min_{M_0 \vartheta} \| M_0\, \boldsymbol{m}(\vartheta) - \boldsymbol{d} \|_2^2. \tag{5.4}$$

If $\hat{\vartheta}$ is a solution of the minimization problem, then it is possible to write:

$$\widehat{M_0} = \arg\min_{M_0,\vartheta} f\left(M_0,\hat{\vartheta}\right) = \frac{\boldsymbol{m}\left(\hat{\vartheta}\right)^H \boldsymbol{d}}{\left\| \boldsymbol{m}\left(\hat{\vartheta}\right) \right\|_2^2}. \tag{5.5}$$

Substituting (5.5) in (5.4), we can obtain an estimate of ϑ independent of the static component M_0:

$$\hat{\vartheta} = \arg\min_{\vartheta} \left(\boldsymbol{I} - \frac{\boldsymbol{m}(\vartheta) \cdot \boldsymbol{m}(\vartheta)^H}{\left\| \boldsymbol{m}(\vartheta) \right\|_2^2} \right) \cdot \boldsymbol{d}, \tag{5.6}$$

where $\boldsymbol{I}$ is the identity matrix. This is a variable projection problem, which can be solved with standard optimization tools. Then, M_0 can be estimated as a second step by applying Equation (5.5).

5.3   MAXIMUM INNER PRODUCT SEARCH (MIPS)

With $\boldsymbol{d}$ as the temporally evolving signal evolution of a voxel, and under the assumption that its noise is Gaussian, the problem in Equation (5.6) is equivalent to a Maximum Likelihood Estimation (MLE). This MLE can be written as a Maximum Inner Product Search (MIPS), looking at the maximum correlation between the measured data, $\boldsymbol{d}$, and the predicted magnetization values, $\boldsymbol{m}(\vartheta)$:

$$\hat{\vartheta} = \arg\max_{\vartheta} \frac{\langle \boldsymbol{d}, \boldsymbol{m}\,(\vartheta) \rangle}{\| \boldsymbol{d} \|_2 \| \boldsymbol{m}(\vartheta) \|_2}. \tag{5.7}$$

In this case, it is relatively easy to formulate the problem as an exhaustive search over a pre-computed dictionary obtained simulating $\boldsymbol{m}(\vartheta)$ for certain values of ϑ, where ϑ is evaluated over a grid of physically acceptable parameters. We can evaluate the scalar product:

$$C_{\vartheta} = \frac{\sqrt{\sum_{j=1}^{N_t} m_j(\vartheta)^{\dagger} d_j}}{\sqrt{\sum_{j=1}^{N_t} d_j^{\dagger} d_j} \sqrt{\sum_{j=1}^{N_t} m_j(\vartheta)^{\dagger} m_j(\vartheta)}} , \qquad (5.8)$$

for each ϑ in our space, where $\dagger$ is the complex conjugate. We then use a lookup table to extract the underlying physical values of the dictionary entry achieving the highest value of C, that is the *highest similarity* or *best match*. Then, M_0 can be estimated as a second step, again by applying Equation (5.5):

$$\widehat{M_0} = \frac{m(\hat{\vartheta})^H d}{\|m(\hat{\vartheta})\|_2^2} . \qquad (5.9)$$

5.4 MRF "MATCHING" WITH MIPS

One method that takes advantage of MIPS to estimate quantitative parameters is MRF [5-1]. In a typical MRF implementation, a *dictionary* is precalculated of the possible transient-state signals using Bloch equation simulations (see Chapter 3 for Bloch equation simulations) over a range of possible, relevant tissue parameters (ρ, T_1, T_2, etc.) and system imperfections (B_0, B_1^+). MRF relies on a transient-state acquisition achieved by the variation of acquisition parameters after each *TR* and can be acquired with undersampled snapshots. As discussed in Chapter 2, there are many different contrast-encoding acquisitions that can be tailored to maximize the contrast between the relevant tissue properties. Once the data are acquired and transformed into the image domain, the aliased signals in individual locations are compared with unaliased dictionary elements (see Figure 5.1).

A simple way to solve the MIPS problem is to perform a *grid search*, i.e., to compute the inner product between the vector representing the acquired data and all dictionary vectors, outputting the indices of the vector with highest inner product. A grid search is the "pattern matching" technique conventionally considered for MRF. Such an approach is straightforward and effective, as it does not require more complicated calculations, such as the computation of gradients of functionals, and can provide global solutions in non-smooth search spaces. However, a grid search is also computationally demanding. A MIPS problem performed on 128×128×128 image voxels (2M voxels) and different simulated parameter sizes with a naive search requires several trillions inner product computations. The estimated computation times are reported in Table 5.1 for various dictionary sizes.

Table 5.1: Benchmark values (time per each pattern matching problem) of 3D MRF on a brain using a 128×128×128 image matrix, given a pre-simulated dictionary, with: (1) only T_1 and T_2; (2) fat fraction estimations; (3) blood perfusion; and (4) a model including the whole water diffusion tensor, currently intractable. All data was simulated with complex singles

	Number of dictionary elements	Number of singular values per element	Total dictionary size	Intel® Xeon® processor E5-2600 v4 (48 cpu)	NVIDIA Tesla K80 GPU
1. MR Relaxation times only	250 thousand	10	2.5 million	615 s	340 s
2. T_1, B_1, B_0 and fat fraction estimations	3.2 million	30	32 million	6.5 h	3 h
3. T_1, T_2, two blood perfusion parameters	300 million	30	3 billion	27 days (estimated)	14 days (estimated)
4. T_1, T_2 and diffusion tensor estimations	15.6 billion	10	156 billion	3.5 years (estimated)	1.5 years (estimated)

To reduce the number of pattern matching operations, several techniques can be used. For instance, methods based on SVD factorization [5-2] or fast group matching [5-3] have been shown to reduce the size of the pattern matching problem without significant loss in accuracy.

The use of a dictionary reduces the requirements to compute all local minima and undersampling effects when compared with a simple least squares fit (see Figure 5.2). This is partially because MIPS searches over a large domain of possible parameters where all minima are considered. MIPS does not require full sampling, as the MRF reconstruction "sees through" aliasing, and aliasing artifacts are "averaged out" with MIPS during the scalar product computation. This effect of "averaging out" can also be observed with reduced aliasing artifacts in estimation using MIPS, where signals of equal amplitude and opposite phase can appear at different timepoints and cancel out in the complex domain, whereas in a nonlinear least squares (NLLS) estimation these errors sum in quadrature, which increases the sensitivity of NLLS to undersampling.

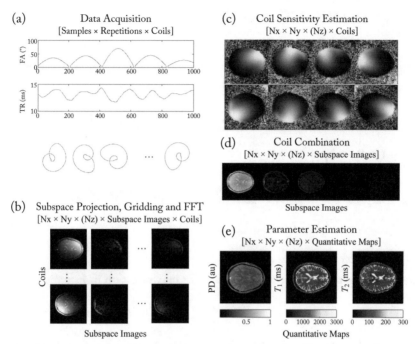

Figure 5.1: In MRF, acquisition parameters are varied at each TR to keep the magnetization in the transient state (a). For fast acquisition, k-space is sampled using non-Cartesian schemes, maximizing sample density in areas rich of signal and undersampling areas that contain lower signal, such as the edges of k-space. In order to reduce the amount of data to be processed, data can be compressed in a temporal subspace using SVD (b). Then, coil sensitivities are estimated in the image domain with adaptive coil combination (c) and coil images are combined (d) for each voxel, the acquired data is compared to the simulation to find a match using pattern recognition (e), generating parameter maps.

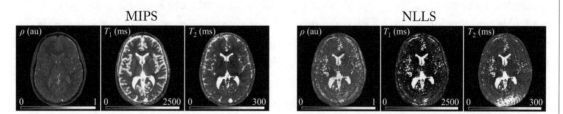

Figure 5.2: Estimated parametric maps from a SSFP spiral MRF acquisition using the method in Figure 5.1 with MIPS, also known as dictionary matching and nonlinear least squares (NLLS). For MIPS, matching noisy data to the simulated dictionary results in consistent parametric maps. Conversely, the high level of noise in the data restricts NLLS fitting and the results converge to inaccurate local minima, even when initialized with MIPS. Images courtesy of P. A. Gómez, TUM Germany.

5.5 UNDERSAMPLING AND MIPS

As proven by successful applications of MRF, MIPS is robust to aliasing due to undersampling. However, if high undersampling factors are used, then significant aliasing artifacts will remain [5-4], [5-5]. Let's review the case of undersampled data in the absence of noise as seen in Chapter 4:

$$d_j(x,y) = \sum_{x'=1}^{N_x}\sum_{y'=1}^{N_y} P^*M_j\left(\theta(x',y')\right). \tag{5.10}$$

By applying $P = (P-P_j) + P_j$:

$$d_j(x,y) = \sum_{x'=1}^{N_x}\sum_{y'=1}^{N_y} P_j^*M_j\left(\theta(x',y')\right) + \sum_{x'=1}^{N_x}\sum_{y'=1}^{N_y}(P-P_j)^*M_j\left(\theta(x',y')\right) = S_j(x,y) + \varepsilon_j(x,y), \tag{5.11}$$

where $\varepsilon_j(x, y)$ is the *undersampling error*.

When computing the MIPS scalar products for maximization in Equation (5.8), we obtain:

$$C^2 \propto \sum_{i=1}^{N_t} S_j{}^\dagger d_j + \sum_{i=1}^{N_t} \varepsilon_j{}^\dagger d_j, \tag{5.12}$$

where "†" is the complex conjugate. We have found that in the presence of undersampling, the value of the MIPS scalar product is biased by undersampling error.

In an ideal world, there would be no undersampling error in the scalar product, in which case the optimum sampling method would disregard these effects and could be obtained by the acquisition schedule maximizing the contrast of signals per each *TR*. However, in practice, the local quantification accuracy in MRF depends both on the used flip angle schedule and on the *k*-space trajectory, as time-dependent point spread functions will have unique interference in different spatio-temporal coordinates [5-6]. There are special cases where the undersampling error has a negligible contribution to the MIPS scalar product. For instance, Equation (5.11) shows that if the evolution of the magnetization, $M_j(\theta(x', y'))$, is sufficiently slow with respect to a full acquisition of *k*-space interleaves, the estimation differences due to aliasing from each acquisition interleave averages out. This behavior is surely part of the reason why most successful MRF implementations have used smoothly varying flip-angle and repetition time schedules.

Undersampling errors from pattern matching on highly undersampled data biases the estimation results in irregular spatial locations. In order to reduce this problem, the time series can be anti-aliased before pattern matching. For instance, a sliding window reconstruction is a simple algorithm that can greatly reduce undersampling artifacts prior to matching [5-7]. For maximum accuracy, the dictionary should be adjusted so that it is equal to the number of *k*-space interleaves used in the window length. This is possible due to the linear nature of the Fourier transformation, which allows the summation of points in the dictionary and in the reconstructed interleaves. Iterative methods such as compressed sensing can also be used as anti-aliasing algorithms. Assuming that the projection into a temporal subspace (e.g., SVD compression) is a fair approximation of the information contained in our signals, the temporal subspace can be used as a sparsifying transform in a compressed sensing algorithm. Written in term of the SVD-compressed vector $\boldsymbol{\alpha}$, as defined in Chapter 4:

$$\underset{\boldsymbol{\alpha}}{\text{argmin}} \ || \mathbf{d} - \boldsymbol{F} \ \boldsymbol{\Phi}_h \boldsymbol{\alpha} ||_2^2 + \lambda R(\boldsymbol{\alpha}), \tag{5.13}$$

where $R(\boldsymbol{\alpha})$ is a regularization functional, corresponding to a local low-rank operator that acts on spatio-temporal image patches [5-8].

The functional $R(\boldsymbol{\alpha})$ is the sparsifying term (see Chapter 4 on compressed sensing) based on temporal compression, where only the first h SVD coefficients are used to represent $\boldsymbol{\alpha}$ providing temporal regularization. In addition to temporal regularization, $R(\boldsymbol{\alpha})$ also produces spatial regularization, following the work on T_2 shuffling by Tamir et al [5-9] :

$$R(\boldsymbol{\alpha}) = \sum_r ||R_r(\boldsymbol{\alpha})||_*. \tag{5.14}$$

The operator R_r extracts a block from each temporal coefficient image centered around voxel $\mathbf{r}$ and reshapes each block into a column of a matrix. The nuclear norm $||\cdot||_*$ is then applied to each matrix and the result is summed.

Such operations can make the representation of the data smoother, allowing methods that rely more explicitly on the computation of gradients (such as simple gradient descent) to better converge to global minima by reducing undersampling effects, as shown in Figure 5.3.

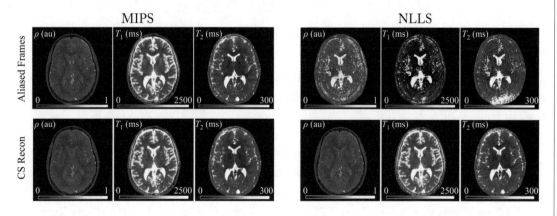

Figure 5.3: Same data as in Figure 5.2. Compressed sensing (CS) reconstruction eliminates aliasing and reduces noise levels, facilitating convergence of NLLS fitting. NLSS results are initialized with MIPS but converge to values that are different from the ones in the precomputed grid of MIPS. Images courtesy of P. A. Gómez, TUM Germany.

5.6   MULTI-COMPONENT ESTIMATION

Under the assumption that one pixel is not a pure tissue species but a mixture, or a linear combination of different species contributing to a portion of the signal, it is possible to write the magnetization in each location (x, y) as the combination of a certain number of signal evolutions $N_{tissues}$, equal to the number of tissues to be modeled:

$$M_j\big(\theta(x,y)\big) = M_0(x,y) \sum_{i=1}^{N_{tissues}} W_i(x,y) \cdot m_i\big(\vartheta(x,y),t_j\big), \tag{5.15}$$

where W_i are the weights of tissue fractions, and where $\sum_{i=1}^{N_{tissues}} W_i = 1$. Note that here each W_i is independent of other tissue parameters such as T_1, T_2 or the static contribution to the signal $M_0(x, y)$. Following this model, if mixtures produce signal evolutions that are sufficiently different from the pure species, it is possible to discriminate tissue fractions uniquely using a combined dictionary (Figure 5.4). An example of this is represented by fat and water separation, possible by including off-resonance in the estimation, and assuming a constant chemical shift of 3.5 ppm between fat and water [5-10].

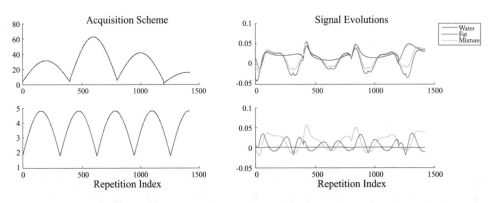

Figure 5.4: Exemplary fat-water mixtures on resonance (100% water, 100% fat, 50% water/50% fat) evolutions: water T_1/off-resonance were set to 900 ms/0 Hz while fat T_1/off-resonance were set to 300 ms/220 Hz.

In addition to fat/water, other multi-component effects can be added to the model in order to achieve more accurate estimations. Other examples of multi-compartment models include brain tissue class segmentation [5-11], as well as myelin water fraction models [5-12].

Other effects that impact T_1 and T_2 quantification are represented by B_1^+ effects. These can be added to the model in the form of a global scaling factor to the nominal flip angle. However, in case of 2D imaging, different locations throughout the slice profile are seen by a different B_1^+, hence a different signal evolution can be written per each slice location, and these can be combined to obtain a single "slice" model [5-13].

5.7   NONLINEAR MR TOMOGRAPHY

Quantitative MRI is normally performed in two steps: the spatial localization to transform time domain signals into spatially resolved images, and the inference from a model for parameter estimation. It is possible to combine the two steps above by formulating the inference problem directly in the frequency domain (see Figure 5.5). This approach combines the spatial encoding and the parameter estimation in a single, large computation, enforcing simultaneously the time-dependent image contrasts with the Bloch Equation, Faraday's law of induction, and spatial-encoding via the accumulating effects of the gradient fields [5-14].

We can start from our encoding equation as in Equation (4.12):

$$s_{j,l} = \sum_{y=1}^{N_y}\sum_{x=1}^{N_x} M_j\left(\theta(x,y)\right)e^{-ik_x(j,l)x}e^{-ik_y(j,l)y} + \sigma_{j,l}. \tag{5.16}$$

We then decompose $M_j\,(\theta(x, y))$ in linear and nonlinear term as in (5.1):

$$s_{j,l} = \sum_{y=1}^{N_y}\sum_{x=1}^{N_x} M_0(x,y)\, m\left(\vartheta(x,y),t_j\right)e^{-ik_x(j,l)x}e^{-ik_y(j,l)y} + \sigma_{j,l}, \tag{5.17}$$

where $s_{j,\,l}$ is modeled as a function both of the space-only components $M_0(x, y)\, m(\vartheta(x, y),t_j)$, and of the Fourier encoding term $e^{-ik_x\,(j,l)\cdot x}\,e^{-ik_y\,(j,l)\cdot y}$, which is independent of tissue parameters. The process $\sigma_{j,l}$ is Gaussian noise. We can now write the full estimation problem directly, without decoupling the Fourier encoding and the Bloch signal model:

$$\left(\widehat{M_0}(x,y),\hat{\vartheta}(x,y)\right)$$
$$= \arg\min_{M_0,\vartheta} \sum_{j=1}^{N_t}\left|\sum_{y=1}^{N_y}\sum_{x=1}^{N_x} M_0(x,y)\, m\left(\vartheta(x,y),t_j\right)e^{-ik_x(j,l)x}e^{-ik_y(j,l)y} - s_{j,l}\right|^2 \tag{5.18}$$

conditional to $m(\vartheta(x, y),t_j\,)$ being a solution of the Bloch equation.

Here, differently from other methods discussed in this chapter, we are looking at solving simultaneously for M_0 and ϑ at all spatial locations, and not at solving each pixel/location independently. The extended set of unknowns calculated by MR-STAT is composed of six unknowns per pixel, and includes amplitude and phase of the static term M_0, as well as T_1, T_2, B_1^+ and off-resonance (ΔB_0) as part of the dynamic component of the magnetization m.

This approach only assumes Gaussian noise on the acquired time-domain data, hence not requiring assumptions on aliasing in the image domain. This is a great advantage over image domain methods, as the assumption of Gaussian noise is more valid in the acquired time domain, as no aliasing is present there even when undersampling. However, this approach also requires a significant amount of computation to estimate parameters. Given that the approach is not local but requires to take into account all spatial coordinates at once, 2D or 3D MR-STAT reconstruction problems are extremely demanding. In a 3D acquisition of 128×128×128 voxel grid, there are 12 M unknowns, since there are 6 parameters per approximately 2 M voxels.

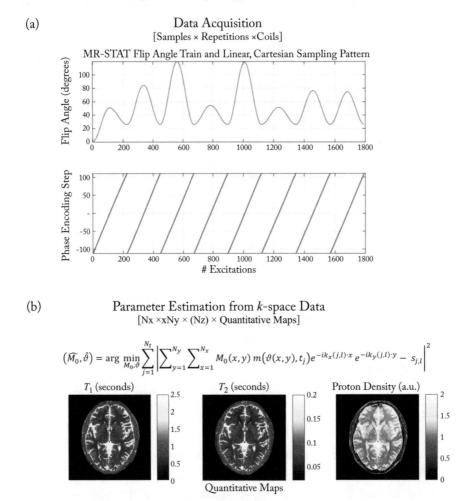

(a)

Data Acquisition

[Samples × Repetitions ×Coils]

MR-STAT Flip Angle Train and Linear, Cartesian Sampling Pattern

(b)

Parameter Estimation from *k*-space Data

[Nx ×xNy × (Nz) × Quantitative Maps]

$$\left(\widehat{M_0}, \hat{\vartheta}\right) = \arg \min_{M_0, \vartheta} \sum_{j=1}^{N_t} \left| \sum_{y=1}^{N_y} \sum_{x=1}^{N_x} M_0(x,y)\, m\big(\vartheta(x,y), t_j\big) e^{-ik_x(j,l)\cdot x}\, e^{-ik_y(j,l)\cdot y} - s_{j,l} \right|^2$$

Quantitative Maps

Figure 5.5: MR-STAT estimates quantitative parameters directly from *k*-space assuming Gaussian noise, hence not requiring any assumption on aliasing structure. Given this property, MR-STAT can be formulated using any *k*-space trajectory, including Cartesian. Representative sequence parameters for MR-STAT are shown in (a), while representative results in vivo are shown in (b). Images courtesy of A. Sbrizzi and O. Van der Heide, UMC Utrecht, The Netherlands.

5.8   MACHINE LEARNING

In the following, we will discuss introductory theory of neural networks (for a more complete introduction see [5-15]) and discuss their use for parameter estimation within quantitative transient-state imaging.

The methods described in this chapter for parameter estimation relied on different approaches for the solution of the quantification problem but shared a common drawback. So far, the methods reviewed had large computational demands (i.e., significant calculation time requirements), which becomes a limiting factor if they are expected to be used within clinical practice. Neural Networks (NNs) present a valid alternative to address these time limitations. The NN approach assumes that parameter estimation can be "learned" by an NN after training on the physical model, such as one generated by a Bloch simulation. NNs have can require significant network training times, but once trained, the NN can greatly reduce storage requirements as well as estimation times, and can maintain high estimation accuracy (see Table 5.2). The computational complexity of NN algorithms at run-time depends on the structure of the network only, and not on the number of samples included in the training. Current literature on "machine learning" methods for transient signals demonstrate models with significantly reduced sizes, which impacts the speed for which parameters can be derived by learning the topology of the Bloch response manifold [5-16].

Table 5.2: Space occupied and estimated execution time for (a) MRF dictionary matching and network for different small dictionaries. Measurement performed on an Intel® Xeon® processor E5-2600 v4 (48 cpu) equipped with a NVIDIA Tesla K80 GPU

	MIPS				Neural Network			
# Words in the dictionary	345	2550	7500	33121	345	2550	7500	33121
Occupied space	8 Mb	60 Mb	173 Mb	763 Mb	1.5 Mb	1.5 Mb	1.5 Mb	1.5 Mb
Time for 256×256 pixels	5s	7s	12s	22s	0.2s	0.2s	0.2s	0.2s

A simple neural network can be thought of as a general linear model, where we wish to determine a function, $f(x)$, that is determined from weighting, w, of each input x to derive a final representation:

$$f\left(\boldsymbol{x}\right)=\sum\nolimits_{i=0} w_i {}^* x_i .$$

(5.19)

This equation, $f(x)$, underlies a "neuron"—which is a single element of a neural network. A neural network is created by the nonlinear combination of a number of these neurons. In general, a neural network tries to find the optimum weights, w, so that $f(x)$ can be determined, given a vector of x. The most basic type of neural network is a feed-forward network, where the linear outputs of each neuron feeds into another layer of neurons. When there are many neurons or layers, this is considered a deep neural network.

However, this model is can represent only systems of linear equations, while we wish to "learn" approximate nonlinear responses, such as those present with transient-state MRI. To achieve nonlinearity, an "activation function" is used: an *activation function*, which we call g, is the function that the linear fit, $f(x)$, in the section above is passed through, resulting in $g(f(x))$. Activation functions are inspired by biology and mirror the firing of a cell after passing an action potential threshold. Activation functions determine whether the outputs of a neuron are amplified or subdued by enabling parameters to increase or decrease in importance during training. A common activation function, for instance, is the sigmoid function:

$$g\big(f(x)\big) = \frac{1}{1 + e^{-f(x)}}. \tag{5.20}$$

The sigmoid function allows the function $f(x)$ to be *activated* or *reversed*. An alternative to a sigmoid activation function, among others, is a *rectified linear unit*, or ReLu. The ReLu is the integration of the Heaviside step function, which is zero for negative values and is linear for positive values. The ReLu is relatively simple to calculate and saves computational processes. The basis of a neural network is the "learning" of the weights, w, of a function, $g(f(x))$, given the input parameters, x, to predict the measured values, y_m. In order to do this, the weights are updated using fitting techniques that usually rely on gradient descent algorithms.

Gradient descent is used to calculate the differences between the measured and predicted values, y_m and $g(f(x))$, respectively, and to iteratively update the weights within the predictive model. Gradient descent can be understood as the least-squares fitting of a linear function, with nonlinear activation, given x inputs and y outputs. In order to evaluate the goodness of such agreement, a Loss function is used. This Loss function depends on the gradient between the estimated values, $g(f(x))$, and the measured values, y. A common Loss Function is $L = E[\Delta(g(x), y_m)]$. The model seeks to minimize the following function $L(m,b) = \frac{1}{N}\sum_{i=1}^{N}\big(y_i - g(x_i)\big)^2$. The Loss function is repeatedly minimized through backpropagation of the derivatives that define gradient descent.

For demonstration, let's state that $g(x)$ is a linear function, $mx + b$:
$L(m,b) = \frac{1}{N}\sum_{i=1}^{N}\big(y_i - (mx_i + b)\big)^2$. The partial derivatives of the gradient inform how we update our estimates of m and b.

$$f'(m,b) = \begin{bmatrix} \dfrac{\partial f}{\partial m} \\[2mm] \dfrac{\partial f}{\partial b} \end{bmatrix} = \begin{bmatrix} \dfrac{-1}{N}\Sigma 2x_i\big(y_i - (mx_i + b)\big) \\[2mm] \dfrac{-1}{N}\Sigma 2\big(y_i - (mx_i + b)\big) \end{bmatrix}. \tag{5.21}$$

Each variable, m and b, is updated based on the calculated partial derivative or gradient between the measured, y, and predicted, f, outputs. Our goal is to minimize any change in f with the variable, such that we come to local minima by ensuring f' approaches zero. In order to do this, we change our estimates of the weights m and b by $-\eta\frac{\partial f}{\partial m}$ and by $-\eta\frac{\partial f}{\partial b}$, respectively, where η is often an arbitrarily determined "learning rate," and the negative is used to cause f to approach zero.

This method can be extended to more variables that are not necessarily within a linear function, provided a derivative is possible. A feed-forward NN can be considered a series of nested equations. $Z(x)= h(g(f(x)))$, where f, g, and h are NNs incorporating activation functions. Then, to find the total derivative with respect to x, we use the chain rule: $Z'(x) = Z'(h) \cdot h'(g) \cdot g'(f) \cdot f'(x)$. This is performed in reverse order from the feed forward order, such that the derivatives "back-propagate" their weights from the final estimation and prediction values. The "Loss" is estimated iteratively for all weights within each function.

As discussed, NNs are very efficient at approximating functions. A continuous, differentiable function can often be represented by a sufficiently deep NN with a finite number of neurons [5-17], it is theoretically possible to train a network to learn the Bloch response manifold as a function of several quantitative parameters, which is difficult due to nonlinearity of the spin dynamic equations. In Figure 5.6, a network with an SVD layer and three NN layers (also called "hidden layers") is trained with Bloch simulated data, and is shown to estimate quantitative measurements in vivo, with results similar to a MIPS algorithm using a grid search.

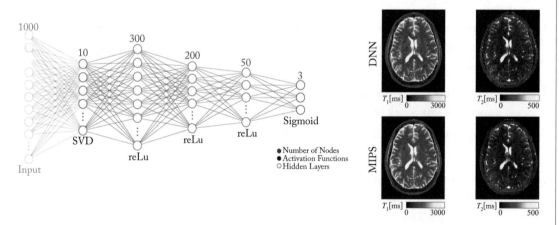

Figure 5.6: Inference of T_1 and T_2 using an NN. Acquisition parameters were based on SSFP MRF and matched the ones in Figure 1: (a) structure of the NN; and (b) the results of the NN estimation corresponded to the ones obtained with MIPS, despite order of magnitude shorter computation times (see Table 5.2). Images courtesy of L. Peretti, University of Pisa, Italy.

5.9 CONCLUSION

In this chapter, we have described several strategies for performing parameter estimations useful for quantitative MRI. In order to bring these techniques successfully to the clinical practice, it is important to visualize the results appropriately and to put these into context of the disease and treatment assessments object of study. In the next chapter, we will discuss visualization, applications and challenges of quantitative MRI.

BIBLIOGRAPHY

[5-1]   D. Ma et al., Magnetic resonance fingerprinting, *Nature*, 495(7440), pp. 187–192, Mar. 2013. DOI: 10.1038/nature11971. 99

[5-2]   D. F. McGivney et al., SVD compression for magnetic resonance fingerprinting in the time domain, *IEEE Trans. Med. Imag.*, 33(12), pp. 2311–2322, 2014. DOI: 10.1109/ TMI.2014.2337321. 100

[5-3]   S. F. Cauley et al., Fast group matching for MR fingerprinting reconstruction, *Magn. Reson. Med.*, 74(2), pp. 523–528, 2015. DOI: 10.1002/mrm.25439. 100

[5-4]   E. Y. Pierre, D. Ma, Y. Chen, C. Badve, and M. A. Griswold, Multiscale reconstruction for MR fingerprinting, *Magn. Reson. Med.*, 75(6), pp. 2481–2492, 2016. DOI: 10.1002/ mrm.25776. 102

[5-5]   J. Assländer, M. A. Cloos, F. Knoll, D. K. Sodickson, J. Hennig, and R. Lattanzi, Low rank alternating direction method of multipliers reconstruction for MR fingerprinting, *Magn. Reson. Med.*, 79(1), pp. 83–96, 2018. DOI: 10.1002/mrm.26639. 102

[5-6]   C. C. Stolk and A. Sbrizzi, Understanding the combined effect of k-space undersampling and transient states excitation in MR Fingerprinting reconstructions, *IEEE Trans. Med. Imag.*, preprint 10.1109/TMI.2019.2900585, 2019. DOI: 10.1109/TMI.2019.2900585. 102

[5-7]   X. Cao et al., Robust sliding-window reconstruction for Accelerating the acquisition of MR fingerprinting, *Magn. Reson. Med.*, 78(4), pp. 1579–1588, 2017. DOI: 10.1002/ mrm.26521. 102

[5-8]   P. A. Gómez, M. Molina-Romero, G. Buonincontri, M. I. Menzel, and B. H. Menze, Designing contrasts for rapid, simultaneous parameter quantification and flow visualization with quantitative transient-state imaging, *Sci. Rep.*, 9(1), p. 8468, 2019. DOI: 10.1038/ s41598-019-44832-w. 103

[5-9]   J. I. Tamir et al., T_2 shuffling: Sharp, multicontrast, volumetric fast spin-echo imaging, *Magn. Reson. Med.*, 77(1), pp. 180–195, Jan. 2017. DOI: 10.1002/mrm.26102. 103

[5-10] M. Cencini, L. Biagi, J. D. Kaggie, R. F. Schulte, M. Tosetti, and G. Buonincontri, Magnetic resonance fingerprinting with dictionary-based fat and water separation (DBFW MRF): A multi-component approach, *Magn. Reson. Med.*, 81(5), pp. 3032–3045, 2019. DOI: 10.1002/mrm.27628. 104

[5-11] C. Liao et al., Detection of lesions in mesial temporal lobe epilepsy by using MR fingerprinting, *Radiology*, 288(3), pp. 804–812, Sep. 2018. DOI: 10.1148/radiol.2018172131. 104

[5-12] Y. Chen, M. H. Chen, K. R. Baluyot, T. M. Potts, J. Jimenez, and W. Lin, MR fingerprinting enables quantitative measures of brain tissue relaxation times and myelin water fraction in the first five years of life, *Neuroimage*, 186, pp. 782–793, 2019. DOI: 10.1016/j.neuroimage.2018.11.038. 104

[5-13] G. Buonincontri and S. J. Sawiak, MR fingerprinting with simultaneous B1 estimation, *Magn. Reson. Med.*, 76(4), pp. 1127–1135, Oct. 2016. DOI: 10.1002/mrm.26009. 104

[5-14] A. Sbrizzi et al., Fast quantitative MRI as a nonlinear tomography problem, *Magn. Reson. Imag.*, 46, pp. 56–63, Feb. 2018. DOI: 10.1016/j.mri.2017.10.015. 105

[5-15] S. Shalev-Shwartz and S. Ben-David, *Understanding Machine Learning: From Theory to Algorithms*. Cambridge University Press, 2014. 107

[5-16] O. Cohen, B. Zhu, and M. S. Rosen, MR fingerprinting Deep RecOnstruction NEtwork (DRONE), *Magn. Reson. Med.*, 80(3), pp. 885–894, 2018. DOI: 10.1002/mrm.27198. 107

[5-17] K. Hornik, M. Stinchcombe, and H. White, Multilayer feedforward networks are universal approximators, *Neur. Netw.*, 2(5), pp. 359–366, 1989. DOI: 10.1016/0893-6080(89)90020-8. 109

CHAPTER 6

Conclusion

Once quantitative maps are obtained, these need to be provided in an appropriate form for clinical evaluation. Repeatable and reproducible quantifications may improve on current diagnosis techniques. In this chapter, we discuss image visualization methods; the development and need for repeatable imaging; and, finally, on the future and need for quantitative imaging.

6.1 QUANTITATIVE IMAGING

Quantitative changes in longitudinal (T_1) and transverse (T_2) relaxation times have been described for many diseases, including neurodegenerative (Alzheimer's [6-1, 6-2], Parkinson's [6-3, 6-4], multiple sclerosis [6-5, 6-6], epilepsy [6-7], schizophrenia [6-8]), autism [6-9, 6-10], oncological [6-11], musculoskeletal [6-12, 6-13], cardiac [6-14], respiratory [6-15], hepatic [6-16], and many other diseases. However, the clinical adoption of quantitative MRI (qMRI) has been limited by the time required to quantify relaxation parameters. Additionally, qMRI is sensitive to MRI system imperfections due to high inter-parameter correlations [6-17, 6-18]. While qMRI was introduced near the inception of MRI, effective qMRI remains challenging and has yet to become universally relevant for clinical disease characterization. Recent advances in qMRI enables it to become more efficient and repeatable. Multi-parametric quantitative (mqMRI) is being studied with new approaches for deriving several image contrasts simultaneously. When quantitative maps exist, synthetic MRI contrasts can then be created from them for advanced visualization of tissues.

6.2 IMAGE VISUALIZATION

In order to enable meaningful radiological evaluation of the quantitative maps, as obtained in Chapter 5, appropriate visualization of this data is required. Contrast-weighted images are the cornerstone of radiological assessments. Contrasts obtained with quantitative imaging protocols can also be useful for diagnosis and treatment assessments. When acquiring dynamic or transient-state signal responses, images for each frame in the transient state are informative of the individual tissue responses. If undersampled timeframes are recovered with anti-aliasing techniques such as compressed sensing, different frames contain different tissue information. This can be done after pattern matching as well, where single frames can be obtained by projecting back quantitative results into the signal domain, i.e., substituting time evolutions in each pixel with the evolution corresponding to the best matching signal, as shown in Figure 6.1.

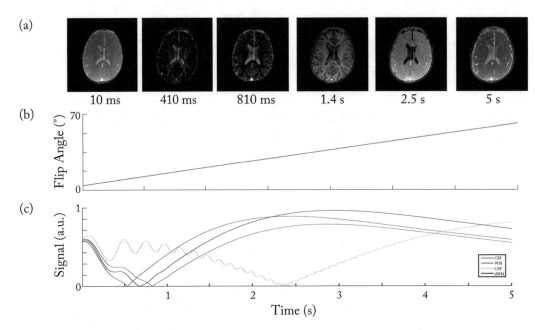

Figure 6.1: (a) Frames from the transient-state response to a flip-angle ramp following an inversion preparation, after pattern matching; (b) the flip-angle ramp used; and (c) the simulated transient-state response of gray matter (GM), white matter (WM), cerebrospinal fluid (CSF), and diseased white matter (dWM).

The underlying T_1, T_2, and M_0 values derived from parameter estimation, such as the ones shown in Figure 6.2, should be useful for improving clinical evaluation. Quantitative values should be comparable across sessions, operators, and scanners, for repeatable and reproducible diagnosis. Some T_1 and T_2 weighted images have reversed contrast compared to quantitative T_1, T_2, and ρ maps. Radiologists are more familiar with weighted images, rather than quantitative maps. The contrasts can either be reversed by changing the color scale, or the inverted map for relaxivities (R_1 = $1/T_1$, $R_2 = 1/T_2$) can be obtained. As radiologists have a limited experience in assessing quantitative maps of T_1, T_2, and ρ, synthetic contrasts for displaying images from standard sequences are obtainable post-hoc. These synthetized images can use sequence parameters matching clinical protocols, or any specific parameters suitable for a particular patient/exam, like shown in Figure 6.3. Image synthesis follows conventional forward models for relevant imaging sequences. For instance, creating a spin echo acquisition with repetition time TR and echo time TE:

$$S_{SE} = M_0 \left[1 - \exp\left(-\frac{TR}{T_1} \right) \right] \exp\left(-\frac{TE}{T_2} \right). \tag{6.1}$$

For a fluid-attenuated inversion recovery with inversion time T_1:

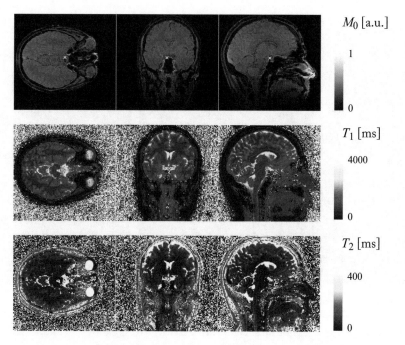

Figure 6.2: M_0, T_1, and T_2 in a healthy volunteer from a 3D MR Fingerprinting acquisition using spiral projections. Like in nuclear imaging visualisations, colormaps can be used to visualize quantitative maps.

$$S_{FLAIR} = M_0 \left[1 - 2\exp\left(-\frac{TI}{T_1} \right) + \exp\left(-\frac{TR}{T_1} \right) \right] \exp\left(-\frac{TE}{T_2} \right). \tag{6.2}$$

And for a spoiled gradient echo with a 90° flip angle:

$$S_{GRE} = M_0 \left[1 - \exp\left(-\frac{TR}{T_1} \right) \right]. \tag{6.3}$$

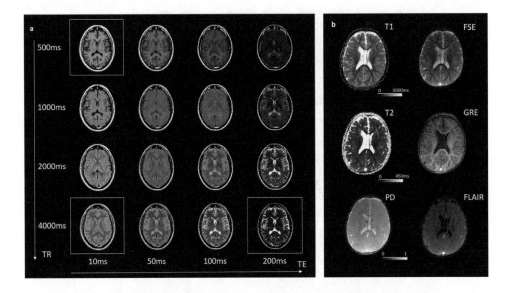

Figure 6.3: (a) Different contrasts obtained a posteriori from QRAPMASTER using Synthetic MR by modifying *TR* and *TE* in a simple spin echo signal model. (b) Quantitative maps and synthetic images in the same subject of Figure 6.1 obtained with spiral SSFP MRF acquisition.

One of the main reasons for the wider usage of contrast-weighted images when compared to quantitative maps is the relatively simpler process to obtain an image. While quantitative maps rely on full modeling and encoding of signals, qualitative contrast-weighted MRI is only accounting for such physical effects in order to produce clinically relevant images. Often, such images are obtained more efficiently than fully-quantitative estimation, and sometimes these are clinically relevant. An example of this is time-of-flight (*TOF*) angiography. *TOF* does not rely on accurate models, but is rather based on the broad concept that fresh, unsaturated spins flow into the slice during a gradient echo acquisition and generate a bright blood signal in the images. Although multi-parametric quantitative imaging is based on the concept of full modeling of the underlying physics, it is possible to add simplified, qualitative models to obtain additional, clinically relevant information from quantitative MRI protocols [6-19]. One example of this is to account for blood inflowing in the slice during the transient-state acquisition, without specifically measuring the velocity, similar to *TOF* angiography, as shown in Figure 6.4.

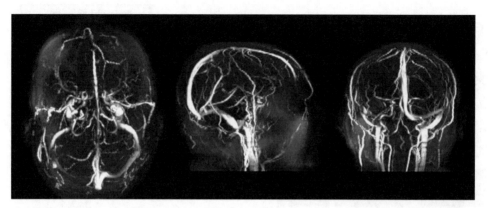

Figure 6.4: Maximum intensity projections from transient-state data using only the 160 frames with the highest flip angle, resulting in angiographic images. This allows visualization of all the main vascular structures in the head, such as the carotid arteries and the superior sagittal sinus. The images show axial, sagittal, and coronal maximum intensity projections of a stack of 3D images (left to right, respectively). Images courtesy of P. A. Gómez, TUM Germany.

During the synthesis of images from quantitative parameters, models used can have different degrees of complexity. It is possible to extend multi-parametric signal models to account for multiple tissues at a single location. For instance, some of the most common confounders for synthetic image evaluations are CSF and flowing blood. To account for these, the dictionary D can be written as a weighted combination of tissue, CSF, and vessels dictionaries (see Figure 6.5) $D_{T,CSF,v}$:

$$D = w_T D_T + w_{CSF} D_{CSF} + w_v D_v ,$$ (6.4)

where $w_{T,CSF,v} \in \mathbb{R}$; $w_T + w_{CSF} + w_v = 1$ are the tissue, CSF, and vessel fraction.

Synthetic FLAIR can then be obtained as:

$$S_{FLAIR} = w_T |PD| \left[1 - 2\exp\left(-\frac{TI}{T_1} \right) + \exp\left(-\frac{TR}{T_1} \right) \right] \exp\left(-\frac{TE}{T_2} \right).$$ (6.5)

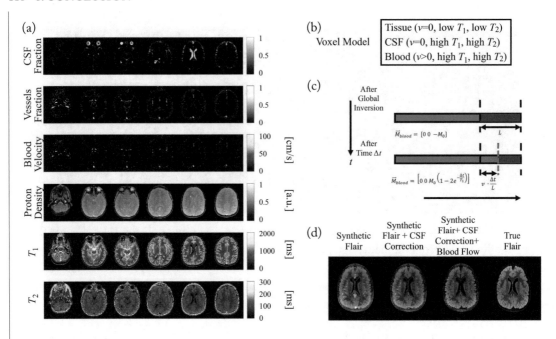

Figure 6.5: (a) Estimations from a three compartment model, summarized in (b). In such a model, soft tissues are characterized by zero velocity and low T_1 and T_2, CSF is characterized by zero velocity and high T_1 and T_2; while blood has a velocity different from zero. (c) Shows the model for blood velocity, where fresh spins, which only underwent the inversion preparation, enter the slice and are excited by the pulses in the transient-state train. (d) The effect of each element in the model on the synthetic FLAIR image, compared to a conventional FLAIR on the same slice. Images courtesy of Matteo Cencini, University of Pisa, Italy.

6.3   REPEATABLE AND REPRODUCIBLE QUANTIFICATIONS

The development of biomarkers includes the "identification of objective and quantifiable medical signs" [6-20]. In order to obtain quantifiable features from images, these should be analysed metrologically, and treated as measurements as in any other science. Mean bias and limits of agreement between different quantification techniques are relevant in order to compare values across measurements. In addition, repeatability and reproducibility are important for individual measurements techniques. Briefly, repeatability refers to the degree of agreement between experiments repeated at the same location, using the same measurement procedure and equipment, performed under similar conditions, and repeated at separate time points. In comparison, reproducibility refers to the degree of agreement between the results of experiments conducted at different locations and with similar but separate instruments. Examples of MRF repeatability and reproducibility can be seen in Figure 6.6.

MRI's sensitivity to many different physical and chemical mechanisms is its blessing and its curse. Difficulties with MRI quantitation thus far have been introduced in Chapters 1 and 2. The large number of degrees of freedom that are available for imaging—both those accessible to a user and the invisible parameters—can often be ignored. We hope that the sensitivity of MRI and understanding its underlying physics will improve, which will guide new technological developments for further repeatable, reproducible research and increased sensitivity to disease states.

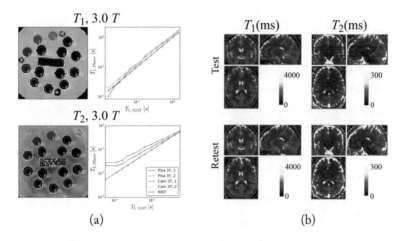

Figure 6.6: (a) Repeatability and reproducibility measurements of 2D spiral FISP MRF in the NIST/ISMRM phantom. (b) Test and re-test in a subject using 2D spiral FISP MRF. Data from Buonincontri et al. 2019 [6-21].

6.4   THE FUTURE OF QUANTITATIVE IMAGING

A move to *quantitative* imaging is being strongly promoted by leading professional societies such as the Radiological Society of North America (RSNA) in their QIBA (Quantitative Imaging of Biomarkers Alliance [6-22]) initiative. In this context, QIBA defined quantitative imaging as "the acquisition, extraction, and characterization of relevant quantifiable features from medical images for research and patient care." Transforming MRI to a *quantitative* science would ideally directly result in enhanced patient healthcare from the accelerated development of diagnostic and therapeutic procedures.

Methods for fast, quantitative, and multi-parametric MRI should continue to develop to obtain similar quality images, but much faster, as current clinical imaging. Quantitative MRI has yet to have had this clear clinical impact beyond cardiac and hepatic diseases. The methods discussed in this book will ideally lead to these developments, which we hope will have a future, if not immediate, clinical impact. The goal is that fast, quantitative MRI will enable better prognosis than is currently achievable for many more diseases, impacting the lives and care of many people worldwide.

BIBLIOGRAPHY

[6-1]   G. Bartzokis et al., In vivo evaluation of brain iron in alzheimer disease using magnetic resonance imaging, *Arch. Gen. Psychiat.*, 57(1), p. 47, Jan. 2000. DOI: 10.1001/archpsyc.57.1.47. 113

[6-2]   X. Tang et al., Magnetic resonance imaging relaxation time in Alzheimer's disease, *Brain Res. Bull.*, 140, pp. 176–189, 2018. DOI: 10.1016/j.brainresbull.2018.05.004. 113

[6-3]   J. Vymazal et al., T1 and T2 in the brain of healthy subjects, patients with Parkinson Disease, and patients with multiple system atrophy: Relation to Iron Content, *Radiology*, 211(2), pp. 489–495, May 1999. DOI: 10.1148/radiology.211.2.r99ma53489. 113

[6-4]   S. Baudrexel et al., Quantitative mapping of T1 and T2* discloses nigral and brainstem pathology in early Parkinson's disease, *Neuroimage*, 51(2), pp. 512–520, 2010. DOI: 10.1016/j.neuroimage.2010.03.005. 113

[6-5]   H. B. W. Larsson et al., Assessment of demyelination, edema, and gliosis byin vivo determination of T1 and T2 in the brain of patients with acute attack of multiple sclerosis, *Magn. Reson. Med.*, 11(3), pp. 337–348, Sep. 1989. DOI: 10.1002/mrm.1910110308. 113

[6-6]   R.-M. Gracien et al., Longitudinal quantitative MRI assessment of cortical damage in multiple sclerosis: A pilot study, *J. Magn. Reson. Imag.*, 46(5), pp. 1485–1490, Nov. 2017. DOI: 10.1002/jmri.25685. 113

[6-7]   A. Pitkanen et al., Severity of hippocampal atrophy correlates with the prolongation of MRI T sub 2 relaxation time in temporal lobe epilepsy but not in Alzheimer's disease, *Neurology*, 46(6), pp. 1724–1730, 1996. DOI: 10.1212/WNL.46.6.1724. 113

[6-8]   P. Williamson and D. Pelz, Frontal, temporal, and striatal proton relaxation times in schizophrenic patients and normal comparison subjects, *Am. J. Psychiat.*, 149(4), p. 549, 1992. DOI: 10.1176/ajp.149.4.549. 113

[6-9]   S. Friedman et al., Regional brain chemical alterations in young children with autism spectrum disorder, *Neurology*, 60(1), pp. 100–107, 2003. DOI: 10.1212/WNL.60.1.100. 113

[6-10]   J. Hendry, T. DeVito, N. Gelman, and M. Densmore, White matter abnormalities in autism detected through transverse relaxation time imaging, *Neuroimage*, 29(4), pp. 1049–1057, 2006. DOI: 10.1016/j.neuroimage.2005.08.039. 113

[6-11] T. Yankeelov, D. Pickens, and R. Price, *Quantitative MRI in Cancer*. CRC Press, 2011. DOI: 10.1201/b11379. 113

[6-12] H. E. Smith et al., Spatial variation in cartilage T2 of the knee, *J. Magn. Reson. Imag.*, 14(1), pp. 50–55, Jul. 2001. DOI: 10.1002/jmri.1150. 113

[6-13] J. E. Kurkijärvi, M. J. Nissi, I. Kiviranta, J. S. Jurvelin, and M. T. Nieminen, Delayed gadolinium-enhanced MRI of cartilage (dGEMRIC) and T2 characteristics of human knee articular cartilage: Topographical variation and relationships to mechanical properties, *Magn. Reson. Med.*, 52(1), pp. 41–46, Jul. 2004. DOI: 10.1002/mrm.20104. 113

[6-14] L. Iles et al., Evaluation of diffuse myocardial fibrosis in heart failure with cardiac magnetic resonance contrast-enhanced T1 mapping, *J. Am. Coll. Cardiol.*, 52(19), pp. 1574–1580, Nov. 2008. DOI: 10.1016/j.jacc.2008.06.049. 113

[6-15] A. Stadler, P. M. Jakob, M. Griswold, M. Barth, and A. A. Bankier, T1 mapping of the entire lung parenchyma: Influence of the respiratory phase in healthy individuals, *J. Magn. Reson. Imag.*, 21(6), pp. 759–764, Jun. 2005. DOI: 10.1002/jmri.20319. 113

[6-16] R. Banerjee et al., Multiparametric magnetic resonance for the non-invasive diagnosis of liver disease, *J. Hepatol.*, 60(1), pp. 69–77, Jan. 2014. DOI: 10.1016/j.jhep.2013.09.002. 113

[6-17] S. Majumdar, S. C. Orphanoudakis, A. Gmitro, M. O'Donnell, and J. C. Gore, Errors in the measurements of T2 using multiple-echo MRI techniques. I. Effects of radiofrequency pulse imperfections, *Magn. Reson. Med.*, 3(3), pp. 397–417, Jun. 1986. DOI: 10.1002/mrm.1910030305. 113

[6-18] S. Deoni, Quantitative relaxometry of the brain, *Top. Magn. Reson. Imag.*, 21(2), p. 101, 2010. DOI: 10.1097/RMR.0b013e31821e56d8. 113

[6-19] P. A. Gómez, M. Molina-Romero, G. Buonincontri, M. I. Menzel, and B. H. Menze, Designing contrasts for rapid, simultaneous parameter quantification and flow visualization with quantitative transient-state imaging, *Sci. Rep.*, 9(1), p. 8468, Dec. 2019. DOI: 10.1038/s41598-019-44832-w. 116

[6-20] K. Strimbu and J. A. Tavel, What are biomarkers?, *Curr. Opin. HIV AIDS*, 5(6), pp. 463–6, Nov. 2010. DOI: 10.1097/COH.0b013e32833ed177. 118

[6-21] G. Buonincontri et al., Multi-site repeatability and reproducibility of MR fingerprinting of the healthy brain at 1.5 and 3.0 T, *Neuroimage*, 195, pp. 362–372, Jul. 2019. DOI: 10.1016/j.neuroimage.2019.03.047. 119

[6-22] A. J. Buckler, L. Bresolin, N. R. Dunnick, D. C. Sullivan, and For the Group, A collaborative enterprise for multi-stakeholder participation in the advancement of quantitative

imaging," *Radiology*, 258(3), pp. 906–914, Mar. 2011. DOI: 10.1148/radiol.10100799. 119

Authors' Biographies

Guido Buonincontri, Ph.D., is a physicist who has been working in the field of MRI for approximately 10 years, alternating between pre-clinical and clinical MRI research. For his Ph.D. program and for a following post-doc he worked on the development and application of novel MRI techniques for cardiovascular assessments in small animals at the University of Cambridge. As a physicist working at the Department of Clinical Neuroscience first and then at the Department of Medicine, his role was to build on the latest MRI techniques to enable studies in basic science and pharmacology, with a main focus on cardiac applications. Currently, he works in Tuscany, Italy, for Fondazione Stella Maris and Fondazione Imago7, two non-profit research foundations part of a children's hospital with close ties to the University of Pisa. Guido has been a Young Investigator Fellow for INFN CSV, as well as a EU Horizon 2020 Marie-Curie Individual Fellow, focussing on fast and quantitative MRI. He is currently conducting an Italian Ministry of Health funded clinical trial assessing fast quantitative MRI in children and challenging adults and is enjoying interdisciplinary and international collaborations.

Joshua D. Kaggie, Ph.D., is a physicist and researcher at the University of Cambridge in the Department of Radiology. Born near Salt Lake City, Utah, U.S., Dr. Kaggie moved to Cambridge in the UK following his Ph.D. Dr. Kaggie did undergraduate research in gamma ray astronomy, which included the manufacture of electronics and the simulation of Cerenkov showers. Dr. Kaggie then moved into Magnetic Resonance Imaging (MRI) research for his Ph.D., where his primary focus was the development of MRI techniques for sodium imaging, which included the development of MRI transmit/receive equipment, simultaneous multinuclear imaging methods, and non-Cartesian acquisition/reconstruction.

At the University of Cambridge, Dr. Kaggie has been primarily funded by GlaxoSmithKline, the Biomedical Research Council, and a European Union Horizon 2020 grant. The collaboration between all three authors began as a result of a Royal Society travel grant, when Dr. Kaggie arrived at Cambridge and as Dr. Buonincontri was leaving Cambridge for Pisa. This collaboration has proven very useful for the development of software and ideas between centers.

Martin Graves is a Consultant Clinical Scientist at Cambridge University Hospitals with over 35 years' experience in both clinical and research aspects of MRI. He received his B.Sc. (1984) and M.Sc. (1987) from the University of London and his Ph.D. (2010) from the University of Cambridge. He is a Fellow of the International Society for Magnetic Resonance in Medicine (ISMRM), the UK Institute of Physics and Engineering in Medicine (IPEM). He has been awarded Honorary Membership of the UK Royal College of Radiologists (RCR) (2016) and the

IPEM Academic Gold Medal (2018). He has co-authored over 200 peer-reviewed publications as well as co-authoring a number of textbooks including *MRI: From Picture to Proton* (CUP 2003, 2007, and 2016), *Physics MCQs for the Part 1 FRCR* (CUP, 2011), and *The Physics and Mathematics of MRI* (Morgan & Claypool Publishers, 2016).

Printed in the United States
by Baker & Taylor Publisher Services